T0331057

A Definitive Guide to Behavioural Safety

'Behavioural safety' or behaviour-based safety (BBS) has been around as a concept for several decades and is commonly held to mean directly tackling the front-line behaviours that lead to incidents and injury. Unfortunately, virulent criticism of some approaches of BBS frequently generalises to all others, so that commentators don't know if they are arguing or agreeing. This book aims to cut through the waffle to be the one-stop guide to all the core theories and principles that underpin behaviour-based safety.

In this second edition, internationally acclaimed behavioural safety expert Tim Marsh leads the reader through the three main strands: The awareness approach, the 'walk-and-talk' approach, and the Six-Sigma safety or the Deming-inspired 'full' approach that covers the systemic approach to safety observation, measurement, intervention, and analysis, but also incorporates emotional intelligence training aimed at enhancing supervisor-worker trust and communication more generally.

Updated to reflect systemic changes due to the COVID-19 pandemic and featuring a brand-new chapter on well-being that discusses the massive changes in thinking about the interaction of culture and personal safety that have occurred since the previous edition was published. This book allows the reader to set up an ambitious and wide-ranging behavioural safety programme from scratch or refresh their current approach.

- Written by one of the world's leading BBS experts, including leading the first major research project on the applied use of BBS outside of the US, based on the author's experiences with more than 400 organisations.
- Delivered in an accessible and popular writing style that simplifies complex theory and arguments in a user-friendly way, plus features end-of-chapter checklists to underpin learning.
- Possesses a post-COVID focus, bringing behavioural safety into the 2020s, and covers the growing concept of well-being in a brand-new chapter.

The title features end-of-chapter checklists to confirm understanding of the concepts. A Definitive Guide to Behavioural Safety is the go-to book for any practicing occupational health and safety professional.

A Definitive Guide to Behavioural Safety

Health and Well-Being

Second Edition

Tim Marsh

CRC Press
Taylor & Francis Group
Boca Raton London New York

CRC Press is an imprint of the
Taylor & Francis Group, an **informa** business

Front cover image: Shutterstock

Second edition published 2025
by CRC Press
6000 Broken Sound Parkway NW, Suite 300, Boca Raton, FL 33487-2742

and by CRC Press
4 Park Square, Milton Park, Abingdon, Oxon, OX14 4RN

CRC Press is an imprint of Taylor & Francis Group, LLC

ISBN: 978-1-032-58420-1 (hbk)
ISBN: 978-1-032-57989-4 (pbk)
ISBN: 978-1-003-44999-7 (ebk)

DOI: 10.1201/9781003449997

Typeset in Times LT Std
by KnowledgeWorks Global Ltd.

Dedication

For my good friends, the inspirational speakers Ian Whittingham (MBE) and Jason Anker (MBE), and the other workers around the world who didn't go home in the same condition in which they arrived at work.

Contents

SECTION I

SECTION II

About the Author

Tim Marsh, Hon Professor at the University of Plymouth, was one of the team leaders of the original UK research into behavioural safety in the early 1990s. He has been a Chartered Psychologist since 1994 and was made a Chartered Fellow of IOSH in 1996, and has worked with more than 500 organisations around the world since, including the European Space Agency and the BBC. He specialises in human error and organisational culture assessment and change. His first major project was with the UK MOD, where he researched recruit suicidal behaviour. For many years, Tim ran the open courses on Behavioural Safety and Safety Culture for IOSH, was awarded a 'President's Commendation' in 2008 by the International Institute of Risk and Safety Management, and was selected to be their first ever 'Specialist Fellow' in 2010. The author of several bestselling books, he has contributed dozens of articles to international magazines, including the *Safety and Health Practitioner* (now IOSH Magazine) and *Health and Safety at Work* (now Sentinel). Tim has a reputation as a lively and engaging speaker and has chaired and presented keynote talks at dozens of conferences around the world. In 2013, he was invited to give the keynote 'Warner Address' at the 60th BOHS International Conference, and in 2016, he was invited to give the closing keynote at the Campbell Institute inaugural 'International Thought Leadership' event.

Anker and Marsh Limited can be contacted via their website: www.ankerandmarsh.com

Preface

'Behavioural safety' or behaviour-based safety (BBS) has been around as a concept for several decades now, following academic studies around the world and ground-breaking commercial work in the United States. It is commonly held to mean directly tackling the front-line behaviours that lead to incidents and injury and is offered commercially under 1001 differing alternatives. These can be boiled down to three main stands: The 'be careful or you could get hurt' awareness raising approach (often delivered by someone who *has* been hurt). The top-down 'walk-and-talk' approach with challenging and coaching at its heart (see STOP), and the Deming-inspired 'full' approach involving workforce participation, analysis, and measurement. (This might be called 'six-sigma safety').

Unfortunately, virulent criticism of some approaches frequently generalises to all others, so that commentators don't know if they are arguing or agreeing, and on many a conference stage have seemed, to this chair, to be doing both simultaneously.

In addition, research and commercial work in Europe, especially has sought to develop the field by incorporating the best concepts from influencing skills and behavioural economics ('nudge' theory) but also process safety models inspired by the work of James Reason and others. For example, Reason's Just Culture and Cheese Models, Sidney Dekker's work on Hindsight Bias, and Andrew Hopkins' 'mindful culture.' The author argues in this book that these concepts really should underpin anything that flies under the BBS flag, with strong criticism of 'be careful out there' behavioural initiatives that they put the onus on the individual where Just Culture shows us that 90% of the root causes tend to be environmental – so analysis and facilitation will always be a far more important component of a methodology than even the most skilled praise and coaching.

The author also makes the case that the principle behind Heinrich's triangle must be a foundation stone of the behavioural approach but, again, the triangle has recently come under sustained attack as a flawed concept because its predictive validity can prove weak in certain situations. This is throwing the baby out with the bathwater – the *principle* (that you cannot guarantee luck either way, but you can significantly impact how much luck you might need) remains one of the most important in all of human endeavour and includes difficult to predict process safety events! Consultants, seeking to sidestep criticisms by marketing 'new' concepts as 'beyond the Triangle' don't help.

Introduction and Context

This new version is intended an overview of behaviour-based safety (BBS) post-COVID. But it's also an attempt to provoke a debate, as I believe BBS has reached a crossroads and requires a major shift in thinking. This is because far too many approaches around the world that fly under the 'BBS' banner either simply exhort workers to 'try harder' and/or use methodologies that only attempt to coach people out of 'bad' behaviours. Alternatively, some others seek to change the person's underlying values so that the same result is achieved.

However, with Just Culture analysis, as described and so well defined by James Reason and Sidney Dekker, showing that typically the vast majority of the cause of unsafe behaviour is *organisational*; such approaches, though often valid, will not, indeed often *cannot*, produce a step change in safety. Therefore, if we are to maximise behaviour change, we must focus on holistic methodologies that first analyze the causes of behaviour so that we can best facilitate changes to the environments that foster it.

It helps that there's a huge amount of common agreement around the work of academics such as Reason, Dekker, and Andrew Hopkins and academics and hugely experienced consultants like Aubrey Daniels, Scott Geller, and Dominic Cooper. However, while there's a lot of excellent writing around key principles to build on but I would like to take the debate further.

Critically, I want to argue that *there is no single definitive BBS approach,* and never can be, as no one size will ever fit all. The best way forward is to select something that's broadly suitable to a relevant situation and then tailor it or design something bespoke to best address the specific situation. This book, therefore, follows the principles of a good job specification that helps accurately select the best candidate. With BBS, as with a job description, there are *essentials* and then merely *desirables,* and there is always a danger of confusing the two. Some academics and consultants, I argue, insist that a desirable is essential, while others, responding to many organisations hope that there's a 'magic bullet' solution, actually ignore an essential or two completely. This book seeks to clarify what they are. Or – since, of course, I may be wrong – week to start a clarifying debate.

Constructive debate is one thing, but pedantic arguing amongst ourselves means that lives, limbs, and eyes are lost. The same is true, though for even less acceptable reasons, when efficient consultancy marketing machines look *primarily* to churn

DOI: 10.1201/9781003449997-1

turnover, hit targets, and generate bonuses. (And, when you consider the root cause of many behavioural safety issues, isn't *that* an irony?!)

As above, the most effective behavioural approaches, this book claims, focus on the environment before the individual. Let's, therefore, stop using ominous and potentially threatening words such as 'behaviour' (as in *you*) and talk instead about 'culture' (as in *us*). In short, rather than lecturing people about their behaviours, we need to build a strong *culture from which* safe behaviours flow naturally.

Rather than exhorting or even encouraging safe behaviour we need to pro-actively *facilitate* it.

I'll state that again as clearly as I can. By far the best way to change behaviour to reduce risk is to *facilitate* it. Specifically, and addressing the current 'does wellbeing have a place in safety?' debate, we achieve situational awareness, focus, discretionary effort, engagement, and empowerment not by demanding or requesting it but, again, by facilitating it.

If we do that effectively, then not only does genuine 'bad behaviour' stand out more; it will also be more legitimate to target it directly with coaching or, if it's 'just' to do so, with discipline or even dismissal. Perception of fairness is essential when we are talking about trust, culture, and wellbeing, and as the book describes how, for many companies, safety is now just a sub-set of wellbeing.

REDEFINING BEHAVIOURAL SAFETY

We know about the roots of behaviourism. It's about Pavlov and Skinner, bells, Russian psychologists, and salivating dogs. It's about pigeons in cages pecking buttons for food. It's about 'operant' and 'classic' *conditioning*. It almost inevitably sounds a bit Orwellian to say to someone: 'I want to talk to you about your *behaviour*' and almost never a good start to a conversation for the person on the receiving end. It very likely sounds ominous and can instinctively trigger the rebel that's in pretty much all of us and put people on the back foot.

Flowing from this, a current problem with BBS is that it has developed a really bad reputation in some quarters. Many unions (see such as the UK's Hazards magazine) are attacking the very notion of BBS in its entirety as the work of 'management lackies' and as too 'person-focused.' Frankly, this is not without good reason as, tone issues aside, there's many a fire-walking course with 'leadership' or 'team-building excellence' crossed out on the side of the box and 'BBS' plastered over the top. Inevitably, the learning point of such courses is: If you can do this, you can do *anything* – so off you go and be safe.

These courses can be great fun and personally motivating. They also have face validity, at a glance, and can look like a relatively cheap magic bullet. (The trouble is, far too many organisations, hard pressed and/or short sighted are in the market for magic bullet solutions). However, face-valid or not, they are most definitely guilty as charged by the unions, focusing almost entirely on the *individual – with behavioural* improvements proving short-term, if at all. In short, I agree with the unions – they're little more than a gimmick.

Instead, any BBS approach should be a sustainable and holistic process, not just an initiative. (No matter how spectacular and fun that initiative is).

Further, I'd like to argue that even acknowledged experts can be too person-focused. A recently published article by the esteemed US professor Scott Geller gives a thorough overview of BBS. Geller agrees that the union position is sound and suggests that 'person-based safety' sets a better tone than behavioural safety and, in doing so, takes a more humanistic, collaborative approach to the subject. Indeed, Geller is reputed to have coined the term BBS and has essentially defined it well as 'focusing on what people do, analyzing why they do it, and then applying a research-supported intervention strategy to improve it.'

His methodologies explicitly seek to win hearts and minds in the long term through intrinsic motivation rather than extrinsic motivators. For example, through the use of good quality coaching not bonuses. It was Geller, for example, who first said so accurately:

Stop calling safety a priority (priorities change and tend to be political) it has to be an embedded core value.

However, the behavioural implications of seminal works such as Reason's 'Cheese Model' and 'Just Culture,' Dekker's 'New Way,' Hopkin's 'Mindful Safety,' and Conklin's 'Pre-Accident Investigation' are, however, not mentioned in his overview at all.

Specifically, Geller says that ' ... an evolution to "People Based Safety" incorporates factors beyond behavioural science to *enhance self-motivated involvement* of the wage workers' (emphasis added).

My issue is that even the very best person-centred approach is, by definition, still, well *too person-centred*!

BBS IN CONTEXT

At this point, it's worth addressing where BBS fits into a genuinely holistic approach to safety. At a very basic level, it may simply be impossible to operate safely because of a lack of time, resources, or equipment. Or it may be that safety management systems are badly designed, badly communicated, impossible to access, contradictory, and so on. Problems here can't be addressed by a basic BBS approach – other than the analysis/learning element that says 'this makes safety difficult and needs changing.' However, any improvements here would certainly fall under the banner 'facilitating safe behaviour' – so certainly still part of a holistic approach to reducing risky behaviour, as I'd like to define it.

This new edition, therefore, has a chapter dedicated to a holistic approach to human error that covers wellbeing, mental health, diversity, and psychological safety. It argues that addressing them well all lead directly to fewer blue lights, less blood spilt, and fewer costly and time-consuming investigations and court cases.

The HSG 48 model of error breaks error down into different types of mistakes and violations, making it clear that many conscious violations are triggered by the culture. Combining this thinking with the notion of a 'Just Culture' of course needs a mention here, as anything that encourages and facilitates the systematic and objective understanding of why behaviours occur is, I argue strongly throughout this book, *the* key element of BBS. Indeed, it is a law of social science that the efficacy of our response, blind luck aside, will have an upper limit set by the objectivity of our understanding.

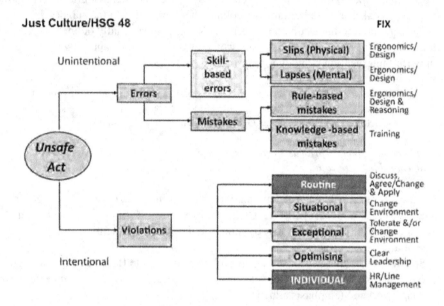

FIGURE 0.1   A Basic Just Culture Model.

In short summary: Unintentional **errors** occur because of physiological limitations such as perception, concentration, memory, or complexity of the task. Anyone who's tried to attempt a crossword when tired knows exactly what this means. A famous example from the world of safety would be the train crash at Ladbroke Grove that led to the second Cullen Inquiry and Report. Initially, the driver was blamed for missing a red light (probably as he was 'not concentrating properly') but when it became clear that perception issues around this light meant that it was the second worst in all the UK for SPADs (signal past at danger), it became clear that the driver may well have been looking straight at it with full concentration but just unable to see it because of on-going visual issues.

And these issues, of course, are often previously reported but at the time not yet actioned. (Learning from and acting on such reports is a key and recurring element of this book).

Consequently, the wellbeing chapter, which goes into depth about issues such as engagement and empowerment, fatigue, menopause, and mental health. All of these, of course, relate to the likelihood of an individual error. Particularly, I hope you might find the pages on 'blue pie slices' representing good and bad days of use and our F.I. scale toolbox talk is (I am assured by many) most probably the best idea we'll ever have. (Sadly, I need to confess, a piece of flippancy that grew legs, not something based on years of research).

VIOLATIONS

Violations are more complex. A violation is essentially where a *conscious* decision to take a risk is made. This might be because of personal need or overconfidence. Often, however, it's because all the more experienced workers act that way so it's 'the way we do things around here and always have' and the individual doesn't want to stand out even on day one despite their misgivings and, potentially, recently received and crystal-clear induction training.

Other violations can be triggered by subconscious cues about expectations. 'I'd like this done safely of course, *'but by Friday please'* means, we all understand, 'by Friday as safely as you can.' This, as we all know that the meat of a sentence is after the 'but' and anything before that mostly waffle. (As is: 'You're a really nice person and I've loved our dates over these last few weeks ... but .... ' They simply don't need to finish the sentence. You know you're going home alone!).

Where an objective analysis of the cause of behaviour can get really problematic is where we cross reference it with blame. I've seen several court cases where a supervisor has effectively said 'but I clearly said safely' and the operative, unable to remember the exact exchange of words/their order/their subconscious understanding replies 'well, yes you did, but I took it that ... er .... ' At this point, in the absence of filmed footage, it's over to the lawyers.

Perhaps *the* major problem here is that we're, as a species, hard wired to blame ...

FIGURE 0.2

THE FUNDAMENTAL ATTRIBUTION ERROR

We are all prone to putting too much emphasis on the person and not enough on the environment, especially when something has gone wrong. It's called the 'Fundamental Attribution Error' (first studied by Lee Ross in 1967). And because it's hard wired and instinctive, it'll automatically kick in unless we make sure we make conscious efforts to stop it doing so.

If you've ever come close to causing an accident on the road by poor driving or lack of attention – you'll probably have been apologetic and aware of all the environmental factors that might have caused it. These might include badly aligned mirrors, fiddling with the sat nav, blind spots, fatigue, being dazzled by a bright light, the children distracting you by squabbling in the back, arguing with your partner, general stress, etc.

But when it's someone else that nearly causes us to have an accident, we will typically be very much more instinctive and hostile. ('Oh, you ***** idiot!' anyone?).

The key thing is that if we were able to talk to them, then often, we'd find that they could, most usually, explain their mistake as we would explain ours. (And on those occasions when we realise it's a neighbour and/or friend, we tend to forgive them *instantly* … 'oh it's only you!'). Again, importantly, in having this conversation about why the error/issue occurred, we might learn something of use to us going forwards.

We really should cross reference here to the hugely influential 'Safety Differently' movement which says 'workers are not the problem, they are the solution' and Poke Yoke management philosophy which says 'blame the process not the person.'

Flowing from this, it's vital to respond to an incident by asking the question 'why?' curiously with the assumption that the response will be at least semi-sensible and one from which we can probably learn. Here we can talk about basic root cause investigation techniques such as 'five whys.' This suggests we keep asking why, if it makes sense to do so, until we come to an obvious root cause. This can take up to 5 times but often doesn't need to.

In simple terms, the five whys trick is to keep asking 'why' curiously and to keep going until we have obviously reached the end of the line and though it can be asked again, it doesn't make much sense to do so. 'Because I simply didn't know that it was risky,' 'because it's much cheaper,' 'because it takes so long to do' would be good examples.

Here's a genuine example of a division of a client that had developed a learning mind-set seeking to learn from events before someone got hurt. You'll see that rail's hazards and near misses exceeded highways by a factor of 250 per employee or so. (As you can imagine the CEO first asked a 'what the hell?' question at the safety forum and when that was answered insisted that highways replicate what rail were doing).

Before moving on, its perhaps worth addressing here the controversy about the safety differently SD movement alluded to above. Some critics (for example the BBS pioneer Dr. Dominic Cooper, author of books such as 'Behavioural Safety' and 'Strategic Safety Culture') have suggested that it's 'more like a cult than a movement' and that critical analysis of it is simply not allowed. However, I'd like to suggest that

Health & Safety Statistics 2008			
	Highways	Rail	Cost
Fatal	0	0	
Major	3	0	136500
3 Day	11	0	60291
Minor	99	0	207603
Hazards & Near Misses	21	683	
AFR (5.45)	3.2	0	
Days Lost	104	0	21736
No. of Employees	1558	202	
Hours Worked	2828907	525134	£426130

FIGURE 0.3 Client data showing the power of near miss reporting.

in truth there need not be any controversy. The basic principles are sound but, it can be argued strongly, simply *not*, in the large, new, or different from previously established best practice. Especially as much of this existing best practice was based on the thinking and earlier works of *Dekker himself*!

For completeness those principles are that: even the best people make mistakes, that's it's often quite easy to predict what those mistakes might be; that learning is essential and that culture is king (a subset of that being that what we praise and don't praise and how we react generally essential. See elsewhere in this book 'everyone contributes to the culture all day every day whether they wont to or not - so they may as well do it positively).

Therefore, I'd strongly argue that models suggesting that it's a step change evolution in safety thinking – implying, therefore, and very importantly, that all other previous methodologies, therefore, *lesser*, are incorrect. (However, though not new, it's thinking that's arguably not anywhere near as widespread around the world as it should be and so broadly, methodologies implemented under the SD banner are really very welcome).

'Stupid' Errors

I'd like now to address those behaviours that look even at *second* glance like they are genuinely individual errors – caused primarily by selfishness, stupidity, or both.

> *I worked with the head of a building company once who challenged me 'this is all well and good Tim – but what about the f***wits?'. This next section outlines the reasoning behind the response I gave him that: 'but we're all of us capable of f***wittery, you, me ... all of us'.*

This is because very often the person involved in blatant 'f***wittery' simply aren't 'stupid' at all. (And even if they are an objective understanding of why they occurred will help reduce blame and point us at an objective and pro-active methodology that might well prevent future f***wittery and/or minimise its consequences in future).

It is central to fully understanding one of the biggest changes in safety thinking in recent decades – the systematic consideration and reduction of hindsight bias. Sidney Dekker, and others, point out that with the benefit of hindsight, we can easily state 'clearly they chose to zig when they should have chosen to zag' putting the blame on the individual for a 'decision-making' error. Dekker stresses accurately that, often, however, if we *genuinely* put ourselves in the shoes of the person in question then, on the day, both zigging and zagging would have looked equally appropriate.

ABC ANALYSIS

At this point, we need to consider human fallibility at its most basic. ABC or 'temptation' analysis, is an explanation as to why people can be tempted to take shortcuts and/or come up with 'work-arounds' that contain more risk than is technically necessary and/or desirable.

ABC stands for antecedents, behaviour, and consequences. Antecedents means the situation and context before the behaviour – which might include access to equipment, risk assessments, tiredness, organisational culture, mind-set, training, induction, and the like. A problem is that experience shows that many organisations, especially those with a largely compliance driven mind-set, give too much emphasis to these antecedents. Classically, they point at a training and/or risk assessment file and feel that by communicating the risk and how to mitigate it, they've 'done their job.'

However, people are not machines, and where that mitigation of risk is in any way difficult or inconvenient, they may give in to the temptation to take a shortcut or develop a 'work-around.' (In everyday life, we might consider risky behaviours such as drug taking, smoking, drinking to excess, eating too much poor-quality food, joining a gym but rarely using it, unprotected sex with relative strangers, or extramarital affairs. Or while driving cars, speeding, 'amber gambling,' driving when you're only 95% certain you haven't drunk too much alcohol and the like).

This temptation arises because it's very often the *consequences* that drive the behaviour. It depends on the behaviour in question, of course, but often, as in the everyday list above, the consequences can have, in practise, far more weight than rules, regulations, and procedures. Consequences can be three things:

- Soon or delayed,
- Positive or negative, and
- Certain or uncertain.

Consequences that are soon and/or positive and/or certain are tempting, especially where two or all three of those apply. (Many readers will recognise this as overlapping greatly with the excellent 'PIC NIC' work by Aubrey Daniels and others. PIC – positive, immediate, and certain. NIC – negative, immediate, and certain).

Hence jumping a red light, when you know there are no cameras and can clearly see there are no other cars coming, saves time and comes with relatively little risk. Smokers say that the first cigarette of the day is just wonderful and know that they might (if they are lucky) not ever succumb to heart disease or cancer – and even if they do, that's probably decades away. Optimists can insert the word 'only' into the widely known sentence 'smoking kills 50% of all smokers' and consider a 50:50 flip of a coin bet something they can live with. (Pun intended!)

In short, everyday examples are legion, and the truth is that, while in an ideal world, such psychology would be left at the 'factory gates' people at work remain, well, just *people* and this basic psychology applies in *all* situations.

Flowing from this, all managers need to be trained in the importance of asking hypothetically, 'is there anything slow, inconvenient, or uncomfortable about doing this job safely?' especially keeping in mind that *psychological* discomfort is often as important as physical discomfort. We know that if the answer to the above question is yes, then there will be a temptation to come up with a shortcut/work-around, and it's just a head count of how many give in to it and how much risk accumulates. (We'll discuss the 'triangle' principle of tending to get the luck that we deserve in detail later).

In reality, we find that many young workers would rather take a risk that makes them a little scared and uncomfortable than risk having more experienced and confident colleagues tease them for being 'timid.' (Please feel free to insert a word instead of timid here that would actually be used in a canteen or the back of a van in your company).

RISK TOLERANCE AND RISK APPETITE

When talking about such 'real world' risk management, it's also well worth remembering that as a species, it has been adaptive for us to be very risk tolerant (Einstein: 'A ship is safest in a harbour, but that's not why we built it'). It's one of the reasons we're such a successful species. (As the oceans start to boil, we're really having to appraise what we mean by 'successful' – but that's a different debate).

However, it is also vital to factor in that our progress (or 'progress') is driven in large part because we have a certain risk *appetite* that we are born with to a greater or lesser extent. (See Dan Peterson's 'Rules for Life'). For example, if young children are provided with a play area that's too safe, they will often adapt it and climb *over* it rather than through it. This is because risk gets the blood pumping so can make us feel 'alive,' much as a horror film or fairground ride can energise us. More than that, showing mastery of risk helps reassure us we have some degree of control over what is often a very scary and hostile world.

These factors, of course, impact hugely on the practical application of risk assessments – especially dynamic risk assessments. Put simply, identifying a risk and then communicating this risk and how to mitigate it does *not* mean that the employee will thereafter automatically avoid it.

Case Study. Formula 1 averaged more than a death a year for decades before the San Marino Grand Prix in 1994. Infamously, the dramatic improvement began there as it is, infamously, when both Ayrton Senna and Roland Ratzenberger died and a third driver, Rubens Barrichello, was badly injured.

However, since 1994 there has been just the one death in 29 years. (The Frenchman Jules Bianchi in Japan in 2014). By any measure that's something of an impressive step change and transpires it came from a combination of both 'top-down' and 'bottom up' methodologies.

The Top-Down Element. Many press conferences were held directly following the crashes by the likes of Max Mosley, Bernie Ecclestone and others. What they said can be summarised as: "This is a personal tragedy for the families of Ayrton and Roland, but also a disaster for Formula 1 itself. Nothing like this can happen again or it could be the end of the sport. So, we insist our world class engineers turn their expertise to safety as well as to speed and, more than that, then share any innovations that they come up with". (Subsequent innovations included deeper tyre walls, longer run off areas, head and neck braces, better PPE, enhanced survival cells, wheel tethers, halos, F1's version of the 'black box' and others).

I'm no motoring expert so can't discuss the merits of each but as mentioned above, the figures speak for themselves.

The 'Bottom Up' Element. What is less well known is that some months later, at the German GP in July, another incident made the front pages when fuel ignited – largely because of the hugely spectacular photographs. (It led to only the driver losing his eyebrows and many a power-point slide about the vital importance of PPE has used it!).

What's most interesting from a cultural perspective, however, is that the investigation showed that, amongst other factors like a fumbled attachment, a fuel filter had been removed by a relatively lowly 'trying to be helpful' mechanic to speed up petrol flow by a fraction of a second. You might ask why anyone would risk something like this to save a few tenths of a second but it's worth remembering that the world record for a pit stop is 1.82 seconds (Red Bull). This ludicrously quick figure did not of course involve refuelling as well as a tyre change but you get the idea. This really isn't a Kwik-Fit 'get yourself a coffee mate and come back in half an hour' situation but one where fractions of a second count. Informed commentators have suggested that in retrospect, it's become clear that understanding the vital importance of engaging with and empowering the most front-line staff proved as important as leadership in the success story.

The key point however is that not one single driver has 'slowed down' by as much as a fraction of a second.

They're just managing the risks far better.

CLASSIC MODELS OF SAFETY AND BBS

To repeat myself slightly, early approaches drew very heavily from the Skinner model of behaviour change, which encourages the use of praise to change behaviours. ('Catch a person doing something right' in Blanchard's 'One Minute Manger' terms). They have also incorporated good quality coaching techniques to engender 'discovered' or intrinsic learning.

However, as above, if 90% of the cause of unsafe behaviour is organisational, then we need to take a more holistic and integrated approach that draws on other disciplines that take that broader perspective. For example, the Australian Andrew Hopkins, a sociologist, has contributed hugely to the world of safety thinking with books on the Texas City and Deepwater Horizon disasters and the 'mindful' safety concept. (Summarised as: 'All organisations are full of problems, the best ones go out proactively to find out what they are, the weaker ones wait for the problems to find them.'

However, James Reason's famous Cheese model, perhaps *the* most important piece of safety thinking of them all, is illustrative of context here.

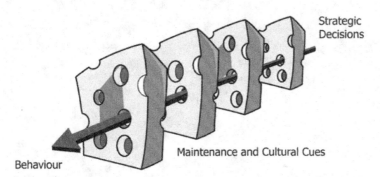

Successive layers of defences, barriers, and safeguards

FIGURE 0.4　Reason's Cheese Model

As above, we know that in the middle of the cheese people may cut corners when told, for example, 'safely but by Friday please,' taking it to mean 'by Friday as safely as you can.' Or when unsafe norms are not challenged by supervisors. Further down, at an individual level, PPE might not be worn because it's uncomfortable or slows the work – or a worker might lose concentration because they are tired or distracted and trip or stray into the line of fire.

Further up the hierarchy, we might consider such contracts between organisations and I'd like to illustrate this with the experience of a client that genuinely meant 'if you can't do it safely, then don't do it.'

When they really delved deeply into the day-to-day operations of one of their new acquisitions (a waste disposal company), they found that they genuinely couldn't both operate safely *and* make a profit. So, they promptly re-sold it. I'd argue that this decision is good – indeed excellent – 'behavioural safety' as I define it. (Though it's worth acknowledging that this is only at an *organisational* level. The individual workers on the lorries still faced the same day-to-day hazards – just in different colour overalls). But ... addressing the political and socio-economic factors needing to be addressed to address this is somewhat beyond the remit of this book).

CONTEXT - CONCLUSION

The key point I'd like to stress is that some BBS methodologies that have been developed over the years reflect, I'm arguing here, a profound failure to understand the complex, holistic, and interlinked reasons why people do what they do. With this in mind, let's now consider the range of methodologies companies have used over the years.

Section I

1 An Attempt at a Taxonomy of Historical Methodologies
An Overview of BBS Approaches

I'd like to suggest that behavioural safety should be grouped into five broad categories.

- **The emotionally resonant 'awareness raising' approach**

  This is a collective expression for such as personal testimony presentations, often from someone who has suffered a life-changing injury. Typically, they will discuss how, if they could have just five seconds back to make a different decision at a given time, they wouldn't have been blinded, paralysed, or maimed.

  The inference is, of course, that if you don't pause and make the right decision when in a similar situation, then *you* too could end up like them. Some of these arguments are extremely articulate, heartfelt, and powerful and will indeed generate a major improvement in safety behaviour. However, their effect is almost always short-term. Typically, even greatly moved people are mindful and act differently for maybe a week before drifting back to their old behaviours. Or perhaps more accurately stated, pulled back by the gravity of the organisational culture.

  These sessions can, however, help kick off a more holistic approach by attracting people's attention and explaining why what's coming next is necessary. Further, they can be invaluable in motivating volunteers to come forward for BBS teams. However, even internationally famous exponents, like my business partner Jason Anker MBE are hugely mindful that these 'be careful or you could end up hurt like me' approaches are, while hugely face-valid and resonant, largely ineffective in the medium to long term. And especially if utilised as a 'magic bullet' stand-alone.

- **The 'STOP' Type Approach (often called 'visible felt leadership')**

  These are based on management pro-actively touring sites looking for workers to talk to about safety, rather than just reactively pulling up workers as and when they see something wrong. Originally developed by DuPont and others, there are 1,001 commercial and in-house variations of this approach around the world.

DOI: 10.1201/9781003449997-3

As they are typically driven through line management, this helps demonstrate a genuine commitment to safety as the workforce can see that there is a clear investment of time and resources. However, their top-down nature means we might often more accurately call this the *'Don't* Approach' with an element of blame implied and too much person-focus.

In transactional psychological terms, it can represent an organisation viewing safety not from an adult-to-adult perspective but from a parental perspective. Transactional Analysis theory argues that each individual interacts with another from one of three positions parent, adult, and child. A child is immature, the adult is rational, and the parent can be either paternal or authoritarian. Ideally, all interactions will be adult-to-adult or inevitably an unintended consequence will follow. For example, working with a 'paternal parent' is a perfectly pleasant interaction but means that the 'parent' is unlikely to be listening to the other person as an equal as the assumption is inevitably that 'they know best!' This will also tend to hinder any empowerment opportunity. Authoritarian parents are tolerated (indeed generally *have* to be tolerated – 'or else!') but discussions about them in the canteen rarely suggest empowerment!

Therefore, although STOP approaches typically demonstrate that an organisation genuinely values safety, they can often prove a counter-productive way to approach the issue. (You, the reader, have most probably responded badly to someone coming at you from the 'critical parent' perspective since about the age of ten and from 'paternal parent' from your late teens/early 20s). The problem is that lots of things that naturally flow from any irritation can weaken a culture. Here we're talking about such as disengagement, indignation, or the withholding of information or discretionary effort.

In short, STOP approaches done well will often achieve good results, but anything delivered from 'parent' is highly likely to also pay a price from an empowerment perspective and, carried out badly, the law of 'unintended consequences' means that they can prove worse than taking no action at all.

- **A 'Nurturing Parent'/Coaching style STOP approach**
  I'd argue that the defining feature of these methods is that they contain an explicit element of objective learning as well as coaching.

  The SUSA approach used by John Ormond in the UK is an excellent example and has nine clear steps encouraging supervisors to use *coaching* and *discovered learning* techniques as well as giving feedback.

  (If you're interested in knowing more about this approach, in the next section of this book, on applied techniques, I attempt to thoroughly analyse and deconstruct each of the 9 steps in turn).

  Nurturing-style STOP approaches are more effective, win awards and feel much more pleasant for a workforce. However, they can be argued not to lead to an apportioning of time and effort in line with what's actually happening on the shop floor and why, as they tend to reactively focus on what can be readily seen and/or comfortably discussed. They will, I argue, therefore normally not prove as effective as something squarely-based on ***pro-active*** and objective analysis and learning.

Consider, for example, asking the question 'Why aren't you following rule X?' in two very different ways. An aggressive use of the query suggests that the listener is a risk-taking fool. A gentler application infers curiosity and concern. The difference totally transforms perceptions, mindsets, and responses. 'Talk to me about the day-to-day realities of rule X' is even better. (Again, the core Safety Differently question is arguably 'I want you to work productively and safely – what do you need from me to do that?' This is a pro-active, open-ended question driven by the workers experience. Even better).

- **'Six Sigma' Safety**
W.E. Deming was an American engineer and statistician credited with inspiring the Japanese post-war industrial boom. He studied systems, their variation, and how the human factor impacts on efficiency and is considered the father of Total Quality Management and Continuous Improvement. Six Sigma is perhaps the best-known current variant of his systemic approaches and was famously endorsed by business leaders including former General Electric chief executive Jack Welch.

Deming's approach crucially involves *analysis* as well as *workforce involvement* and, commonly, *measurement* too. (So, delivering the benefits of 'what gets measured gets done' and 'if we can measure it, we can manage it'). Applied to BBS, it can therefore add goal-setting sessions and feedback charts to the methodological options. (With Tom Krause and BST perhaps the most well-known examples of pioneering work here).

The upside is that, resourced well, they are, of course, very effective. The downside is that, like almost all continuous improvement processes, they need a lot of on-going resourcing. (Again, please see later material in this book for detailed descriptions of the methodologies involved).

- **A cultural or holistic approach**
Based squarely on the Just Culture model, I believe this is the best behavioural safety approach. As above, the Just Culture model works on the principle that the reason for an unsafe act is typically environmental, not personal, about 90% of the time. Therefore, logic dictates we should spend around 90% of our time and resources *analysing and improving the environment in which we want behaviour to be safe* and just 10% looking at the individuals concerned. The principles of learning teams (see Kaizan and later references to BRCA or behavioural root cause analysis teams) and HOP principles very much overlapping these last two categories.

From this perspective, *measurement* and *feedback* then become *desirable* elements to be used if appropriate. However, *analysis* is always *essential* as the analysis drives the responses and facilitation efforts that will be most effective. The 2012 London Olympics would give us a simple example. Understanding that many accidents seemed to be caused by fatigue in workers setting off from digs all over London very early in the morning, cheap good quality porridge was made available. (Especially in the light of the worker deaths controversies about Qatar 2022, it's worth remembering that London remains the only ever fatality free major build for this type of event).

INCLUDING ADVANCES IN THE UNDERSTANDING OF MENTAL HEALTH IN A GENUINELY HOLISTIC METHODOLOGY

This topic is covered in depth in the wellbeing chapters but in summary: In the UK, we lose 31 workers to suicide for every one we kill in an industrial accident, with a ratio of around 8 to 1 even in industries such as construction. Further, it's estimated that 1 in 5 workers are suffering poor mental health at any one time and, post the COVID pandemic, those figures are getting worse. (Industries such as maritime quote as many as 1 in 4).

Though data is scarce and complicated by the fact that some people have learnt that if they are feeling depressed, this can be helped by pro-social acts and anxious people can be risk averse, early studies suggest that, on balance, a struggling worker is about twice as likely to injure themselves as one enjoying good mental health. Though research at this point is largely tangential, there are three reasons primarily:

- One, they are more likely to be distracted and low in situational awareness and fail to see risk.
- Two, they are more likely to be fatalistic and indifferent to risk.
- Three, they are more likely to have poor interactions, make bad decisions, and actually create risk.

Research on all those factors being causal in incidents is both plentiful and unequivocal.

A quote from leading UK barrister Laura Thomas illustrates:

In almost every court case I've been involved in following a major incident the person in question was having a bad day in one way or another.

I'd argue that this isn't a political or philosophical view, it's just cold logic. Once we accept that facilitation of safe cultures and environments is key, as well as investment in the design and management of safety, we can, for example, get creative and can step into the world of behavioural economics (AKA 'Nudge Theory' beloved of so many governments). In addition, it's natural to consider wellbeing, fatigue management, and free porridge as part of a holistic approach to facilitating and enabling safe behaviour.

Though some behaviour may need to be addressed directly, especially if it presents a clear and present danger, primarily we must treat it as a sign that something is wrong with the organisation, not the person. (Again, the Poke Yoke mindset is key. If there are any assumptions to be made, it's that it's the organisation not the person that needs changing).

HOLISTIC BUT FLEXIBLE

The five approaches I have presented above are (I suggest) suggested to be in order of their typical effectiveness and sophistication. I further argue that a dialogue based on 'what do we need to do?' questioning is often actually far less onerous to resource and run than a complex 'six-sigma' style approach and is equally applicable to peripatetic workers.

However, flexibility is needed and an organisation will want to cherry pick the most appropriate methodology for itself *at the time in question.*

For example, I have few kind words to say about an authoritarian top-down approach, but this is in part driven by a personal loathing of being treated that way. However, the classic situational leadership model by Hersey and Blanchard insists that, in certain circumstances, a top-down authoritarian leadership style is *exactly* what is required. For example, it's best to be directive with staff where the safety implications are severe and/or where staff lack competence and experience. Similarly, an old-fashioned emotional-awareness-raising session might be just the ticket to kick off a safety programme, as long as it's an early part of a holistic approach.

As above, it's unlikely to achieve anything much in the medium to long term on its own, but it does help motivate people to listen to an explanation of what we are going to do and why we're going to do it. If the presentation starts with, or is based on, something upsetting or dramatic, we'll at least have people's full attention. If they are moved and engaged – they are more likely to volunteer. (It is of course what they do after they've volunteered that makes all the difference in the medium to long term).

Finally, it may be that some form of all-singing, all-dancing gold standard approach might simply be beyond an organisation at that time, no matter how motivated to improve they are.

MORE CONTEXT – THE IMPORTANCE OF CULTURE AND THE LIMITATIONS OF TRAINING

Perhaps the key element of stressing a holistic contextual approach to behavioural improvement is that 'culture is king.' We know that the first thing a new start or sub-contractor will do is look to the confident and experienced colleagues to 'see how things are done around here.' Culture being, in essence, not what we say it is – but the day-to-day behaviours that are undertaken, especially when it's busy or no one is watching.

There's a fabled tipping point (see Malcolm Gladwell's book of that name) where if a certain number of people are doing something, then even strong minded and independent people will feel pretty much compelled to follow suit. (There's a PhD thesis or three begging to be written about where it is for key safety behaviours generally and in specific situations. I think around 80 or 90% myself and that 50:50 essentially gives those new starts and subbies free choice).

Here's a simple example. Imagine you pick a hire car in a new country and are warned that speeding is treated harshly. As you leave the airport, you see the sign in the picture. You'd certainly be very well primed to stick to the 80 kph limit. However, as you join the motorway, you find that the inside lane is averaging 90, the middle lane 100, and the outside lane is a real free for all. I used to ask the question at conferences, 'what speed would you be doing?' But, of course, people responded, 'but you have to keep up with the

FIGURE 1.1

flow for safety reasons.' So learned to ask instead, 'how many of you would have used the outside lane within an hour or two?' Typically, 50% raise their hands!

For this reason, cultures can be self-sustaining as new starts and sub-contractors look around and are influenced by the norms and especially by anyone who looks experienced and/or confident and/or senior. Imagine three new apprentices starting work together. The most confident of the three will, de facto, be a safety leader on day one as the other two will be influenced by them. We often say *'everyone* is a safety leader all day every day whether they want to be or not' (shy new starts on their first few days excepted). The often unspoken concept of a 'taboo' is relevant here. For example, in rugby crowds aggressive and anti-social behaviour simply isn't tolerated even though half the crowd will be drunk. People acting up are policed by their fellow supporters and arrests are very few and far between. Being unfriendly to visiting fans, wherever they are from, simply isn't done. It's taboo. There's great influence in a respected peer asking in a van, factory floor or canteen 'what on earth do you think you're doing?'

One of the golden pillars of transformational leadership is remembering this. (Along with communicate clearly and with impact, empower with dialogue, praise rather than criticise and coach rather than tell). Here's an example of a man, just about to become UK Prime Minister, whom the CEO let onto a Glaswegian ship-building site without PPE because 'not much is going on and the workers will understand my position.'

FIGURE 1.2

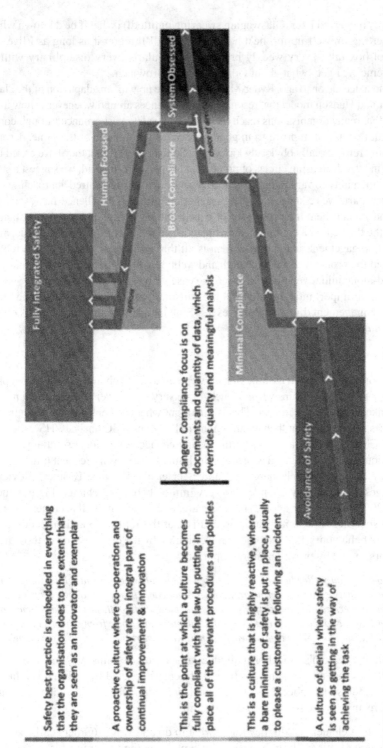

Fully Integrated Safety

Human Focused

System Obsessed

Broad Compliance

choice of err

uptime

Minimal Compliance

Avoidance of Safety

Danger: Compliance focus is on documents and quantity of data, which overrides quality and meaningful analysis

Safety best practice is embedded in everything that the organisation does to the extent that they are seen as an innovator and exemplar

A proactive culture where co-operation and ownership of safety are an integral part of continual improvement & innovation

This is the point at which a culture becomes fully compliant with the law by putting in place all of the relevant procedures and policies

This is a culture that is highly reactive, where a bare minimum of safety is put in place, usually to please a customer or following an incident

A culture of denial where safety is seen as getting in the way of achieving the task

FIGURE 1.3 Ryder-Marsh model.

What (Vince) said later (Glaswegian swearing omitted!) is that if he'd known what he was setting himself up for, he'd have held firm. 'I'll regret it as long as I live … every tool box talk, every weekly brief and particularly every disciplinary with a safety theme … I get "what about Gordon bloody Brown eh?!"'

Around a decade or so ago, Ryder-Marsh staff came up with an adaptation of the classic Parker and Hudson model that resonates with audiences around whenever I show it. It suggests that many organisations reach their best ever safety performance through compliance but then try to improve from good to excellent, with more of the same, despite diminishing returns really obviously kicking in. So many report that they have ended up spinning in ever decreasing circles of process and paperwork. Instead, we suggest a step change in mentality towards a humanistic and genuinely person-centred approach.

In large part, we're suggesting that a generic culture of excellence has excellent safety and mental health as a sub-set of excellence generally. This book constantly refers to the theme that a commitment to objective analysis, constructive dialogue, and (therefore) engagement and empowerment is all that's needed. (And this focus really helps avoid the problem of mental health and wellbeing being treated – ineffectually – as a stand-alone initiative, much has been the case with safety over the years).

Even more importantly, building on the truism 'compliance is but base one,' the degree of engagement directly addresses a second truism: 'Compliance itself is often *discretionary*.'

CULTURE AND TRAINING

Often, an organisation will assume that all we need to do to change is to train people.

This short section explains why so much (often very expensive) safety training might fail to make a meaningful impact. This is important when we consider why even training courses full of excellent 'behavioural' material, such as ABC analysis, HSG 48, and Reason's Cheese model, fail to make the lasting change hoped for. Essentially, many organisations simply struggle to appreciate the limitations of even excellent training.

We say training 'merely base one' and Beer et al. (Harvard Business Review) have suggested that it may often lead to very limited behaviour change. Figures such as 'only 20% efficacy' are quoted, but these are, of course, catch all averages and not situation specific. However, what is agreed is that the follow up and embedding of the desired behaviours, and the practical facilitation by the organisation of those new behaviours, typically prove far more important.

> I must quote Scott Geller here as he came up with the funniest, most insightful and most 'European' safety quip I've ever heard from an American. He says, as above, that education and (proper) training are simply not the same thing and, to illustrate, asks parents to consider how they'd react if there 17-year-old daughter came home from school saying 'we did sex education today' but then added, 'and next week, we're moving on to sex training'.

I think that Vroom's model of individual motivation explains why this is so very clearly. It says that any person's likelihood of being impacted by training is a factor of three elements multiplied. Note that the *multiplication* element is vital as it means a low score anywhere gives a low score overall.

- The first element is 'do they know what to do and why?' (Others call a key element of this 'selling the vision' by explaining the reasoning).

- The second, do they feel they can do what is required? (As well as understanding the training, this covering time and tools, for example).
- The third asks 'is it considered valuable to the organisation?'

Vroom

Why
for example:
– Heinrich
– Just Culture

Basic/generic skills
for example:
– Assertion
– Ice Breaking
– Presentation Skills

  X   X

What
for example:
– Analysis
– Communication

More advanced /specific skills
for example:
– Five Whys Analysis
– Coaching

Embedding the new behaviour
for example:
– Day to Day Feedback
– Formal Appraisal Items

FIGURE 1.4

For some individuals, the third element is sufficiently covered if they *as a person* truly value and understand the 'why' in element one. For many others, however, the response of the organisation is utterly key, and this is largely a cultural piece. For these people, it is vital that they are praised if they undertake the new behaviour and/or criticised if they don't.

A simple case example is where staff are trained that they should always ask for a 'safety assist' if they feel them necessary. In the case study I have in mind, the training around this topic was very well illustrated with several practical and local examples of where an assist may be required.

However, in practise, the feedback from line management was often that workers were asking for 'too many' (and this sometimes verbalised in quite direct language and not just implied). This inevitably inhibited the number they actually asked for. (I was involved in the legal case following a worker who was killed having 'failed' to request an assist and had proceeded alone).

Again, it's worth articulating this thinking as it's often said, 'if they'd had proper training this wouldn't have happened,' but that is far too simplistic an assumption. Indeed, often, and illustrating the simplistic thinking, when there's been a behavioural incident, the action that flows from the analysis is that everyone should go through the training again … (This often referred to as "name, blame, shame - retrain"!)

CONSIDERING HR AND HSE

As wellbeing increases in prominence in the safety world, the interaction with the HR department has become an increasingly hot topic at conferences.

The position of this book is that the more effectively they interact, the better. Two illustrations:

Firstly, failing to make the most of a company's human capital must be at the top of any risk register and any company should of course be set up and organised to mitigate its key risks *regardless* of its structure, the name or number of departments it has, or how they are required to co-ordinate with each other. It is simply an eternal truth that the more integrated, coordinated, and holistic an approach to any risk mitigation, the better. In short, it doesn't matter what department you work for, what your job title is, or what your budget is. We're talking about mental and physical health here, so playing 'silly buggers' politics about budgets and 'standing' is *immoral*, I argue strongly, as we as unprofessional.

Secondly, despite the warning above, we know that a company isn't going to develop a stronger and more sustainable culture without some training.

To develop this to a logical conclusion: Sparkling, happy sheets are great but someone needs to undertake the on-going, on-site coaching of delegates. Safety professionals and safety representatives can be key to this. (Though many companies don't do this at all). But many of the skills, mindsets, and methodologies requested on the course will be generic (e.g., assertion, feedback, listening skills, etc.) and so ideally also be properly embedded in the general *appraisal system*. So, this clearly means safety working with HR. (Then there's the application of a Just Culture decision tree when something's gone seriously wrong. This, of course, *definitely* needs departments co-operating and co-ordinating).

A KEY PRINCIPLE – BEHAVIOUR AND THE NATURE OF LUCK

This short section addresses Heinrich's Principle and explores the idea that even 'simple' behavioural errors are just as much a cultural issue as any other. Further, that the basic model of sound systems, pro-active objective learning, and empowering the workforce always applies – even for 'simple' slips and falls.

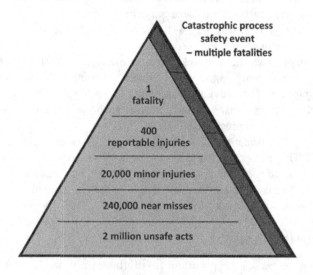

Catastrophic process
safety event
– multiple fatalities

1
fatality

400
reportable injuries

20,000 minor injuries

240,000 near misses

2 million unsafe acts

FIGURE 1.5

Two illustrations as to just how important the issue is. First, almost everyone in the safety world knows that the Piper Alpha disaster claimed the lives of 167 men, but fewer know that, over the years, more off-shore workers in the North Sea have been killed in falls than in all the process safety issues combined. (There are lots of stairs – they are steep, metal, and they are often wet and slippery). Second, statistically, gravity is just about the most dangerous thing on the planet. (It's very difficult to name a famous person who has made the papers for having a serious accident where the expression 'fall' doesn't feature. Think Christopher Reeve, Rod Hull, Ozzie Osbourne, Eric Clapton's daughter, Michael Schumacher ...).

The statistics are simple. If falling down stairs off-shore, steep and slippery though they may be, is only a one in a hundred thousand chance – then two things might flow from this. The first is that a lucky person might enjoy an entire career without ever holding the handrail but never falling. The second consequence is that if approached by a supervisor or colleague, the individual can quote that 100,000 to 1 figure as justification for a robust and/or dismissive response! (And/or the supervisor can quote it to themselves when worried about such a response and seeking a reason to 'blind eye' and walk on past).

However, if overall, the stairs on a platform are used a million times a year, then you're looking at ten falls (give or take) if no one holds the handrail. We should see one fall a year, on average, if 90% hold the rail – but only one every ten years or so if 99% do. And, once we're up in the 90 percent's – and *only* when we're up the 90s for such a statistically key behaviour – does 'zero harm' become genuinely viable. (And that 90% figure is also broadly where the fabled 'tipping point' is where new starts and sub-contractors really feel compelled to follow local norms).

But even for such simple behaviours as holding the handrail, we still need a systemic and analysis-based approach with the safety hierarchy in mind. A case study:

> *I once worked with an HSE inspector who'd been asked to look into a fatal fall at a nightclub. The man who fell was drunk but also truculent and they wanted his view on how likely a fall was. (They were concerned that a doorman, annoyed by his attitude, might have pushed the man down the stairs). What he found was that the stairs were badly lit, steep, warn and slippery (even before adding slopped beer), had steps of differing sizes promoting the odd 'air' step even by sober people and that the handrails were so small as to be merely decorative. 'In truth I'm very surprised this is the first serious fall on these stairs' was his verdict. And of-course it wasn't even close to the first. Indeed, there had been another fatal fall just the year before though this 'entirely accidental' one to a (they presumed) drunken customer was seen by other customers so although thought sad and unfortunate not at all controversial. Nothing to learn here!*

In short, getting the basics right is always step one and, in this example, the club were about as far from a 'culture of care' for their customers as it's possible to get. We always have to pro-actively seek to set people up to succeed rather than fail. However, although we never want to put the onus too much on the individual – the individual is nearly always the last domino to fall, and so the last chance to prevent an injury *is* often in their own hands.

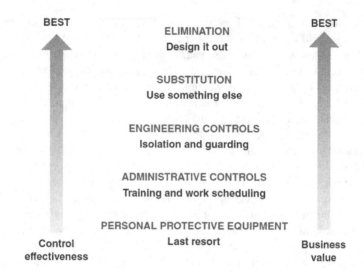

FIGURE 1.6

So, despite all that I've said above about context and culture being king, we *can* develop habits of personal risk management that will stand us in good stead inside or outside the workplace. So now it's set properly in context, we can now address some basic but sound 'behavioural safety' that focuses on the person.

INDIVIDUAL HABITS AND MIND-SETS

I have a video of a motorcycle crash where a car stuck in a traffic jam tries to escape by performing a U-turn but drives directly into the path of an on-coming motorcy-clist. While the motorcyclist is compliant (not speeding and has every right to use the outside lane), he is not pro-active in his riding. (Two pro-active rules of thumb many motorcyclists will be aware of: One, 'always give yourself the time and space to deal not just with your own mistakes but the mistakes of others' and two, 'assume all other road users are drunken idiots').

I once showed this film with an ex-special forces soldier in the room who commented:

You all know me ... someone with very high levels of risk tolerance and you'd be mortified at some of the things I do on a motorbike in the woods. However, this accident wouldn't have happened to me as whilst I'm hugely risk tolerant, I'm also very risk aware. Whilst still thinking about sport or shopping I'd have noticed the curve in the road reducing vis-ibility to no more than 30 yards ahead, noticed the traffic jam full of frustrated drivers and noticed that there was no barrier keeping them from me ... and, instinctively, I'd have slowed and moved to the inside lane to give myself a little more time and space.

This is an excellent example of ingrained habits that can save your life. However, it remains true that an organisation should always, as above, design out the risk as far up the safety hierarchy as possible. In this incident, this regularly congested bit of road really needed a barrier to make sure no one could give in to the temptation to

U-turn especially as the curve in the road meant it was almost impossible to undertake such a turn with any certainty of safety nor was it at all difficult to predict that motorists would be tempted to do so. But, having (yet again) made the 'design out the risk' point, let's return to individual mindset.

Risk and Attention

Studies show that the vast majority of accidents occur *not* when we are doing something highly dangerous as we tend to be alert and/or applying all necessary safeguards no matter what. Here's an example from aviation. It's said that no matter how tired they are, pilots do not ever tend to crash on landing (because of fatigue). Instead, they let their families know they have landed safely then fall asleep at the wheel 50 kilometres into their 100-kilometre drive home.

It's when we are doing something only moderately dangerous and it interacts with one of the following, we tend to have a problem.

- We have habituated to the true extent of the risk (this a risk that can be accelerated if confident experienced co-workers look comfortable).
- We are rushing (either because we are):
  - Time poor
  - Angry and/or frustrated
- We are tired.
- We are distracted:
  - Mental health issues
  - Personal issues
  - Environmental issues (for example noise – or perhaps most (in)famously the 'hello boys' bra advert!).

In such situations, we can put ourselves or others at risk when we lose control of the situation and/or unknowingly step into the line of fire. Training employees in these simple taxonomies so that red warning lights automatically flash (or, more accurately, are more likely to flash) when something above occurs maximises the chance of them stepping back and seeing the risk more clearly.

Tool box talks help but design is, as ever, key. For example, keeping the staff that secures people into harnesses on 'high ropes' courses at adventure parks 'fresh' and alert by regular job rotation would be a good example of best practise here. Once trained, it's a very simple and routine task but it remains utterly safety critical.

Keeping Control

The safety campaigner Ian Whittigham MBE talked about a simple fall to illustrate the importance of not losing control. He'd ask what happens when we fall over, and the most frequent answer is 'nothing.' But sometimes we land awkwardly and break a wrist or a shoulder. Sometimes we suffer life-changing injuries to the head or back. And sometimes we can hit our heads and die. (The actress Natalie Richardson – Liam Neeson's wife – died in such a fall when skiing very slowly on a nursery slope, for example). Ian's point is that once we start to fall, the dice are rolling and we have little, if any, control over what happens next.

The trick is to **keep control** and not fall.

For many of us, driving on the roads is the most dangerous thing we do. For example, every time you drive along a motorway you will see someone breaking the 'two-second rule' as in 'only a fool breaks the two-second rule' (Though please note that 2 seconds isn't anywhere enough in bad weather or if you're tired!). Even in good weather, the driver is in no position to react if anything happens in front of them no matter how fast their reflexes. However, speed awareness courses use data and case studies of horrendous incidents to illustrate that it's almost all about the speed and distance and only a very little about any individual's reflexes. These courses usually leave attendees feeling very uncomfortable in the future when they realise that they have lost concentration and are driving too close to the car in front. (Indeed, not just uncomfortable but genuinely *alarmed* when they realise they've lost focus and it's a large lorry or bus they can't even see past to get any advanced warning ...).

But this awareness of risk needs practise and repetition because the majority of our thinking is *subconscious*. First an example and then the science.

Habituation, Complacency and Your 'Reticular Activating System'

Here is an experiment for you to undertake. Go to a park when there is a stiff breeze and listen to the rustle of the leaves for a while. Then start a conversation or read something. Can you still hear the leaves? You'll find that you can't, and that's your reticular activating system (RAS) allowing you to concentrate on the task at hand by filtering out background noise. The same is true of ticking clocks and the like and, more importantly for this chapter, the same is true of risk perception with many an 'old-timer' enjoying much sport teasing a new start about their fear of some risk the old-timer long ago habituated to.

This is why I strongly push back on the term 'complacency' and instead prefer 'habituation' when talking about apparent over confidence around risk. The trick is to keep reminding yourself 'I may have done this task 1,001 times without hurting myself but it's still as risky as when I first did it.' Then behave accordingly!

The Science

Daniel Kahneman's book 'Thinking fast and slow' is the only psychology book that has won a Nobel prize and makes the case that the vast majority of our thinking – even vitally important thinking – is undertaken largely subconsciously and certainly no-where near as consciously as we'd imagine. Perhaps the classic safety example is driving home and making 101 life-saving risk assessments as you go. Then taking your coat off unable to remember anything about the journey at all.

In this specific context, the benefits of developing good habits so that we *instinctively* trigger if, as above, we drift to less than 2 seconds away from another vehicle are crystal clear. It also explains why the former soldier on his motorbike was probably correct when he asserted that he'd have 'instinctively ...' acted.

So good habits, well ingrained, are very useful to any individual. However, I'd argue that the very essence of a culture is the daily dialogue between any worker and their line manager. So, now we need to talk about safety leadership because ingrained good habits of leadership nearly always effect *many* employees.

2 BBS and Leadership
Perhaps the Single Most Important Contextual Issue

First, we need to talk briefly about Sigmund Freud and what his psychology has to say about safety leadership. He is the grandfather of 'mental gymnastics' and he explained how very few admit, even to themselves, to having a bad attitude towards safety or anything else. Simply telling them they have one will rarely get us anywhere.

In practice, when we are told we have a bad attitude, particularly about something perceived to be as important as safety, Freudian concepts like denial, distortion, rationalisation, and projection may well apply. Good safety coaching and leadership is in essence, then, ideally, a discussion that leads the person to conclude *themselves* that their behaviours and beliefs are misaligned. (Again, telling them they are misaligned almost never works – as they will be 'different'/'experienced' and one way or another immune to the risk with normal rules not needing to apply to them. Again, they simply need to come to the conclusion that they are mistaken themselves!).

The positive side of 'Cognitive Dissonance' (which we are discussing here) is that we generally don't like our behaviours and attitudes to be out of line either way. So, addressing the issue from the other side, if we can get people to behave safely, then often the attitude will follow. Classically, when we persuade a borderline employee to get actively engaged in some aspect of the safety process that they find rewarding, they'll often up and announce they have always been fully committed to safety.

And it's that word commitment that's always key.

LEADERSHIP – A CASE STUDY

Early in my career – i.e., in the previous century! – I spoke at a conference on 'Readiness for BBS' and took two clients with me as case studies. The first group, from an oil company, spoke about how the roll-out of an empowerment programme had primed them perfectly for the sophisticated process we implemented. While the first client spoke, the second set of clients, from a pallet repair outfit, were smiling and laughing.

The first question when they took their turn was of course to ask what was so amusing about the oil company presentation. They explained that they were looking at the readiness slide the oil company referred to and felt that they didn't even qualify for rung one, let alone the penultimate one that the previous speakers felt they'd achieved.

DOI: 10.1201/9781003449997-4   **29**

Their story was a simple one. Their CEO and managing director had been to a funeral in the West of England to bury an employee who had been killed at work. Understandably devastated, they had stopped at a service station on their way back north and made a plan. It included deciding to never promote anyone who didn't hold a basic UK-based safety qualification and also to hire me, who they had come across at a London conference a few months before. I naively proposed a comprehensive BBS approach with safety committees writing their own items, time-intensive measurement schedules providing data to be used at workforce run goal-setting sessions, and charts updated weekly. (Hugely influenced of course by the six-sigma style BBS primarily being used in the USA with success). The committees went for five categories (housekeeping, PPE, man-machine interface, etc. each containing six items; a real logistical stretch for a company with some really aggressive production targets.

There was, of course, strong kickback from management at some sites, questioning whether the leadership really wanted this approach, but that didn't last beyond discovering that they really *did*. Management, led by Neil S and Vince M, ably supported by Dave F and Wayne B among others, drove it through. (Forgive me the name checks – I'll be grateful to these men forever – they helped launch my now 30-year career and find professional meaning in life. Without their genuine commitment, my inexperience would quite possibly have led to the bullish and comprehensive methodologies I'd proposed falling flat on their face!) As it was, a couple of years after the pilot site launched, they were able to show, nationwide, a reduction in accidents to one-tenth of the original levels. It made a stunning case study included in several Institute of Occupational Safety and Health (IOSH) publications.

Incidentally, the oil company was indeed primed for great success and the division we were working with won all sorts of awards. Then another, less positive, learning event: Despite the success, the central office imposed a company-wide scheme based on a simple, largely top-down methodology (as described above) which proved a perfect illustration of how not to do BBS as it instantly undercut all the ownership and empowerment we'd built up over a few years. In this case, it led to a severely demotivated volunteer team and the end of the highly successful scheme we'd help set up.

ONE SIZE DOES NOT FIT ALL

These two contrasting case studies are a reminder and illustration that every organisation is different and that each one needs a tailored approach.

James Reason concludes 'The Human Contribution' with the observation that 'safety is a guerrilla war, and one that you will probably lose eventually.' We must fight a clever, tailored, rear-guard action as best we can for as long as we can. This is the essence of the reality of BBS and applies to individuals managing the risk in their lives. He talks about the overstretched elastic band model – suggesting that in this imperfect world, there is very likely to be real tension in the system and that the nice plans in the file will be put under strain. Here, the knot is in the middle but under strain and inevitably, from time to time and in place to place that the knot will be pulled out of balance by raw material prices, sub-contractors that talk a great fight but then don't deliver. Or it could be new legislation, new competition or weather issues.

Organisations are, to an extent, defined by the speed with which they spot that they've drifted and the speed with which they snap back into genuine balance. The various behavioural safety methodologies described here can really help with that.

FIGURE 2.1   Tim illustrating Reason's 'knot in the elastic band' model with a rope at a conference in Turkey.

However, the one thing all the best guerilla armies have is excellent leaders. It's almost impossible to drive through medium to long-term improvement without senior management commitment. However, we need to help them tailor a methodology that maximises learning and workforce empowerment. Like a good guerilla army, we need to be astute, focused, and flexible. We need to know what we're trying to achieve, why we're trying to achieve it, and what the strengths and weaknesses of our options are. We need robust but flexible plans with options and leeway built in. Operational leaders, like everyone else in organisations, are hard-wired to do what they think will be best-received. They are stuck in the middle, passing on problems that have been passed down to them. The term 'supervisor squeeze' covers it well.

So, at this point, let's address what excellent day-to-day behavioural safety leadership looks like.

LEARNING

Clearly, this chapter suggests that the number one skill a safety leader needs to possess is an excellent approach to learning that focuses on empowering and enabling safe behaviour. For example, re-organising a job so that items don't have to be carried up and down stairs. Replacing harnesses that are difficult to use or fixing long reported leaks that make surfaces slippery. Impactful signs reminding people to 'hold the handrail' help of course, but less so if the signs are old and tired. But, even sparkling new fabulously produced ones are far less impactful than making the handrail easy to hold.

As ever, the starting point is to objectively understand *why* unsafe acts or conditions are occurring before responding and allocating resources. The mechanics of the various learning methodologies are covered in detail in the next chapter but to continue with this more contextual chapter I'd like to address empowerment and engagement. Vital for reducing error or course as switched on and engaged workers make fewer mistakes. (They also enjoy better wellbeing – and please do refer to this later section for a full consideration of the interlink).

Transformational Leadership

This is, in essence, all about communicating clearly, communicating impactfully (using stories and memorable illustration, etc.), coaching not telling, praising not criticising, and knowing that we're all leading by example all the time whether we want to be or not – so we may as well do it well.

To return to the easy to hold handrails. The C-suite always holding them is probably second only in importance as making them easy to hold!

In terms of interactions, after listening positivity and praise is key. 'Catching a person doing something safe' and praising them for that is proven to be about 10 to 20 times as effective at promoting behaviour change as criticising them.

However, I'd argue, neither is as effective as an experienced and respected foreman always clipping on at height. It also makes any censure fairer and nothing undermines trust faster than a perception of unfairness. (Trust is perhaps the key metric for an organisations empowerment level and many experts from the field of wellbeing and stress, such as Sir Cary Cooper argue that the level of trust is the most important organisational metric of them all).

Finally, good coaching is about treating people like adults and using data to illustrate a point. So, to return to a previous simple example, you could say to someone on an off-shore rig.

> *Did you know that the chance of falling if you're not holding the handrail is about 100,000 to 1 so if you're lucky you can work an entire career never holding the handrail and never falling ... but ... the stairs on this platform are used about a million times a year so if none of us hold the handrail we'll have about 10 accidents a year give or take, if 90% hold the handrails then an accident once a year give or take ... but if 99% hold the handrail then we'll only have one incident every ten years or so ... and whilst we're talking did you know that more people have been killed in falls off shore than in all the process safety accidents combine. Piper Alpha included ...?*

In short, the most effective cultural and behavioural techniques and principles apply as much to trips and falls as to any other safety issue. Studies of coaching show that what we're really trying to achieve is a 'light-bulb' moment where the person 'gets it' and thinks 'OK, I do actually see what you mean' because that thought is full of internal *volition* and therefore tends to stick.

CONCLUSION

It's worth returning to perhaps the most simple taxonomy of 'great leadership' and then considering, in summary, how it applies to generating day-to-day safety behaviour.

- Sell the vision – explain what we need to do and why in a clear compelling way.
- Provide the tools.
- Provide on-going monitoring and feedback to embed.

Good quality coaching is just explaining the 'why' in a powerful and impactful way. (Or giving negative feedback in a constructive way). We must expand 'provide the tools' to include not just the right spanner and enough time to do the task – but also to the culture that sets excellent norms and expectations and high levels of analysis, empowerment, and engagement.

It's easy to say 'in this company of "brother's keepers" we challenge, we welcome challenge and we speak up.' But words are cheap. World class behavioural safety is where the commitment is there to use the various methodologies available to proactively facilitate and enable these behaviours and mindsets on a par with the way quality and speed of production inevitably are.

3 Wellbeing, Mental Health, Diversity, and Psychological Safety

INTRODUCTION

In recent years, we've had a movement that has been summarised as 'we must stop shouting safety and only whispering health.' The figures are staggering. The BOHS suggests that for every person killed in the UK by an industrial accident, around 100 will die a premature death because of exposure at work, and the Rushton report actually suggests this is a 'best case' figure.

This shift has quickly grown to incorporate mental health, as well as exposure to toxins as, in addition, and central to *this* chapter, in the UK around 31 working age people will take their own lives for every life lost to an accident. (UK ONS data suggest more than 4,000 working suicides per year compared to less than 200 fatal accidents). Driven to an extent initially by London-based construction companies and Australian initiatives (or so it felt), mental health was already receiving much more attention prior to COVID and is absolutely front and centre now we've emerged from the COVID pandemic.

Encouragingly, however, there's also a growing understanding that mental health is central to a win-win approach to empowerment, engagement, excellence, and risk management generally.

Issues that were controversial even a few years ago now simply aren't, and this chapter seeks to summarise why so many companies now refer to wellbeing, health, and safety *in that order*. It seeks to explain, for a largely safety orientated audience, why wellbeing excellence has a huge impact on safety excellence as well as on a range of key KPI's such as turnover (especially of best staff), absenteeism, presenteeism ('quiet quitting'), discretionary effort, and excellence of performance.

Two 'truths' being stated at conferences with, in my experience, no hint of push back. One is that a wellbeing strategy must be something that is developed and reported at the board level as a key element of organisational sustainability and excellence. (Not at a lower, tactical level or you'll likely end up only with an EA programme hardly anyone contacts and 'bikes and bananas' initiatives whose primary benefit seems to be free fruit and subsidised gym membership for those employees who already eat healthily and take exercise!).

The second issue is that for every Euro spent, the rate of return (ROR) is at least 5–6 Euros, which of course reflects and is entirely congruent with the KPIs listed above. A 2024 UK based study by Deloittes finding that for every pound invested in wellbeing a company will get around £4.52 back). Suddenly, in this time of global financial crisis and staff shortages, an excellent wellbeing strategy

DOI: 10.1201/9781003449997-5

isn't a nice to have. Increasingly, it's being seen by even the most hard-nosed business types as a key element of business sustainability. Get it wrong, and the organisation's very viability may well be compromised.

The good news is that the world of safety has long known what is required for excellence. Now is the time to help organisations apply that hard-won knowledge and experience to what is, it is very easy to argue, a considerably bigger and inter-related risk.

A CLASSIC 5 FACTOR HOLISTIC FRAMEWORK OF WELLBEING

There are many wellbeing models, but most conference papers allude to this five-factor version.

- Good work;
- Good health;
- Financial Security;
- Good Relationships;
- Giving

This is primarily a business book, of course, so this chapter focuses very much on why good work is good for you and how. The other factors will be touched on, however, as taking a genuinely holistic and integrated approach is essential.

Good Work

It helps if you have a job that you enjoy because good work is good for you (but, of course, bad work certainly isn't).

Multi-million-selling writers such as Marcus Buckingham emphasise that we are unlikely to excel at anything unless we enjoy doing it, as otherwise we simply won't put in the energy, effort, and passion to maximise our potential. Of course, economic realities can crash against Hollywood 101 'follow your dream' films here, and there are exceptions, though sometimes they illustrate the principle.

For example, reading John McEnroe's biography, it's striking that he didn't actually enjoy his tennis career. He was driven mostly by a fear of failure. He certainly excelled, but this internal conflict definitely showed sometimes! He now has a job that he genuinely does enjoy, commentating, and seems a different, far happier, person entirely.

Organisations can, of course, help with this element by maximising involvement and empowerment while being mindful not to over-empower to a level that is stressful. Peter Warr's vitamin model of mental health at work is decades old but helps explain that and is still an excellent framework for understanding whether an employee makes their way home from work with a spring in their step … or not.

Warr's factors are:

1. Money: Relating to the base of Maslow's hierarchy of needs and also the second of the five main factors considered in detail next. In brief, it reflects the fact that we all have to eat, and most are driven to provide a safe and

comfortable home for ourselves and our families. (Importantly, though not having enough is hugely stressful, studies suggest that once we have 'enough,' then the marginal benefit of more is limited).

2. Physical security: In some respects, this is an obvious one, also reflecting the Maslow hierarchy's most basic need.

   By this, Warr doesn't just mean not being involved in an accident or developing a bad back, but also not being in conditions that are physically stressful such as heat and noise. The world of safety has long known that in the middle of a busy shift, a worker will do many things to make life more comfortable, such as not wearing a hood that's heavy, cumbersome, and fogs up so you can't see what you're doing. It's annoying and slows you down. Failing to wear it might cost you your eyesight, your health, or even your life, but very probably not today.

   These are the two most obvious and generic elements. The next 7 are more psychological and though some are more interested in money than others and risk appetites vary (see above) more likely to vary person to person.

3. Opportunity for interpersonal contact: Some will want a lot and some a little, but we all want it to be good quality. We'll return to this, as other issues covered in brief here, in detail later in the chapter.

4. (and 5) Opportunity for skill use and variety: Self-evident elements, with obvious links to meaning, achievement, and flow. ('Flow' is where you get absorbed in what you are doing – especially where it's challenging but you are doing it well – and time feels like it is passing quickly).

   Though we say 'self-evident' a recent example from the UK shows how a lack of these elements, even on a specific day, can have severe consequences. After the Manchester bomb attack of May 2017, fire services personnel were, as detailed in a report published in November 2022, unnecessarily held back from attending to victims for more than two hours as the scene had been over-cautiously declared too hazardous to attend as the attack might be on-going. In May 2018, the fire service reported a surge in stress related absence and take up of counselling services in the previous year as their firefighters struggled with feelings of 'guilt' and frustration that they could have helped but were held back.

6. Valued social position: (Again, see Maslow's hierarchy of needs). It's a primeval drive to be considered useful, so we're not likely to be pushed out of the cave/away from the camp fire any time soon. That basic drive aside, it's a very personal thing, and we might – very briefly – consider economic migration as it illustrates the individual perspective. Some people in the west may look at a relatively menial job and think, 'I'm not doing *that*!', even if it's all they are qualified for. Others, coming from a different mindset, will instead think, 'A job! A real, paying job. Excellent!' and they'll stride from their front door with energy and pride, even at 6 a.m. Naturally, employers respond to these attitudes and the behaviours that flow from them.

7. Opportunity for control: Some people are control freaks and hate being told what to do by anyone or anything ever. They even swear at their sat nav.

Others have a greater tolerance and prefer little direction. Indeed, they may find too much autonomy positively stressful. For example, people who have chosen to work in uniforms can be quite low in this need, and people who choose to be self-employed can be quite high.

A practical day-to-day example of best practice in safety and health: A tool-box talk about the day's tasks should be 25% dialogue *at least*, allowing anyone who would like to say something a genuine opportunity to do so. Deny them that opportunity with a rushed 'any questions or concerns?' with voice tone and body language that clearly means 'you'd better not have as we need to crack on' and the person will actually leave the meeting even less motivated than when they arrived.

8. Goal and task demands. Ideally hard but realistic and clearly understood and communicated by someone who has explained 'why' as well as what.

A key model here is that of optimal pressure. If goals and task demands are too high, then we'll of course, be likely to be stressed and can suffer 'burnout,' but if they are not high enough, then we are likely to be a little listless, demotivated, and performing far below our potential (some call this 'loafing' or 'rusting up').

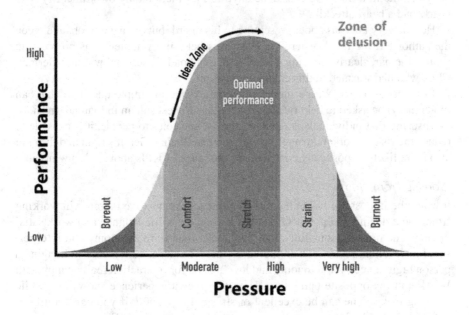

FIGURE 3.1   The Stress Curve.

Task identity and traction is important here, and links to 'meaning' generally. It reflects how well we understand how what we're doing fits in with the general picture. A famous example would be at NASA, when someone sweeping up told the visiting John F. Kennedy, in response to the question 'What are you doing?' replied 'Helping put a man on the moon, Mr President.'

Again, toolbox talks and frequent communication with supervision is key here.

9. Level of uncertainty: We don't want to be told 'can't say' when asking about rumours, we don't want to be wondering if our infamously moody manager will be in a good or bad mood today, and we don't want to be that forklift truck driver working flat out in a blame culture who knows that one thing they really need to avoid the sack is an extended run of good luck as they strive for '... but *by Friday*.' We don't want to be at an airport or train station and told 'your flight/train is delayed' but nothing else!

Frankly, it's relatively easy to describe what good work is made up of. However, ensuring we maximise and tailor work for each person we work with (because every colleague is first and foremost a person) requires thought, co-ordination, and effort.

'Flow', Meaning and Safety Projects

Being involved in a successful process or project, HS&W-related or otherwise, can be hugely rewarding and energising. For example, I once had a safety team volunteer say, 'Twenty years I've been with this company, and last night (preparing a presentation) was the first time I've ever done anything for them on my own time. My wife nearly had a heart attack!'

He was very nervous about presenting to his board, but did a good job, and when the author left the hotel, where the event was held, the volunteer was propped up against the bar, clearly enjoying a bit of a 'boozy bonding session' with the German CEO, who had warmed to this chap's enthusiasm.

In practical terms, Warr's model suggests a range of simple questions that can (potentially) be asked to help tweak a role or task to best suit an individual. In short, by ensuring that individuals are suited to roles, and roles to people, it is possible to reduce the risk of workers struggling with their wellbeing. Get this right and they are also more likely to positively contribute to the range of KPIs above. It's a win-win.

Working From Home

It would be impossible to not mention the impact the massive increase in working from home during and post-COVID has had on work-life balance and work satisfaction. The most obvious links with the above issues are autonomy and interpersonal interaction. Again, it's an incredibly personal issue as it's easy to imagine a person high in need for autonomy and low in need for interaction really happy with WFH, and the opposite being true. However, recent experience shows that while working from home can be excellent or stressful especially if you have family or not ... it can actually vary not just day to day *by the hour!* (For example, being home to wave the children off to school at 8 am is wonderful but having them back at 3 less so ... etc.).

Being able to have a lot of meetings in the comfort of your living room is convenient, but the split-second delay in transmission is shown to hinder real rapport and hence can be stressful. Worse, we know the meetings simply aren't as 'rich' and that those spontaneous 'watercooler' moments are impossible. (Recent evidence suggests that not being physically present can lead to an element of 'out of sight, out of mind' and if management feel that if you're simply a resource on a screen, then that screen can be anywhere in the world and possibly resourced far cheaper than you!).

Likewise, being able to answer e-mails while in the meeting is excellent if it's a bit you really don't need to be paying attention to. Rather less so if you do. (And don't even start me on training courses where delegates are allowed to switch off their cameras entirely so don't even need to pretend to be listening!)

FINANCIAL WELLBEING

Some statistics. The Rowntree Foundation says that 13 million working age people in the UK are living in poverty in 2023. The charity Shelter suggests that typically, one in three employees are one pay check away from not being able to pay their rent. In addition, one in three low earners regularly borrow to cover their rent, and 150 families in Britain become homeless every day.

And, as above, all around the world following COVID and the wars in Ukraine and the Middle East, things are getting worse.

On the other hand, as above, a recent Kronos study shows that, once we are earning above the amount required to live on the threshold, doubling your income increases a sense of wellbeing by a score of just 0.2 (on a 10-point scale). While, remuneration did make the top 10 reasons for employees leaving their job – it came just 9th. The relationship with money isn't even entirely linear as, for example, a surprisingly large percentage of people (30%) who win the lottery say it made them *less* happy in time as they handled the *change* badly.

These stats are very illustrative that money isn't happiness and satisfaction for most people, but it is, of course, important that they have *enough* money and the better we do with the other four elements discussed here, the less we need! For example, if we give workers a 20% pay rise out of the blue, job satisfaction goes through the roof as you can well imagine. For one or two months. Then it goes back to pretty much where it was before!

Books such as 'The Spirit Level' argue that absolute wealth isn't as important to happiness as relative wealth compared with others – which is why observations such as 'We're all in it together' work less well when delivered by multi-millionaires! The internet doesn't help here as it makes comparing ourselves to others easier and there isn't a wellbeing guru on the planet who doesn't advise 'don't compare.'

Even people who think they have this in balance can be very wrong and the following quotation often applies:

> Many of us are stressing ourselves doing things we don't enjoy, earning money we don't really need, to spend on things we don't really want, to impress people we don't even like …

However, the sobering statistics above illustrate that a significant percentage of the working population are just getting by from day-to-day. As well as impacting directly on mental wellbeing, such immediate and significant personal concerns are very likely to impact on attention and concentration, and could certainly therefore also contribute to accidents in the workplace.

Recognising that money worries can impact severely on stress levels and work performance, several organisations around the world have started to offer basic courses on 'financial wellbeing' as almost all of us could do with being more financially astute. And for those who are 'useless with money' and struggling, it has to be better than learning lessons from lenders with exorbitant interest rates or the local loan shark! Interestingly, companies running such courses often report the same two things. One, more people turned up to the session than expected, and two, more people tick the 'that was really useful' box than expected.

So, if your organisation has yet to provide them, they really should start.

FAMILY AND FRIENDS

This is a huge and personal area of wellbeing that interacts with all the others in a dynamic way. I can't possibly do it justice here, but this is a holistic piece, and there are some comments worth making for orientation. These are:

All non-technical skills for maximising the empowerment and potential of your colleagues see later in the chapter – of course apply equally to family and friends. (Indeed, perhaps *the* guru of positive psychology, Martin Seligman, says that we should have a ratio of 3:1 praise to constructive criticism at work, but 5:1 at home. This translates as: You should always be your child's biggest supporter, but not to the extent that you ruin them).

As ever, the Heinrich principle applies here too. (This that you tend to get the luck that you deserve depending on the effort you put in – though there are no guarantees either way).

Every day, people are walking in the countryside, talking and laughing with friends or family, and 'splashing out' by buying themselves a (now) 'earned' treat of a choc ice or a coffee halfway around the walk. We all know how good these things taste in circumstances like that. Huge numbers of readers will have promised themselves they will find time for 'far more of this next year.' (Especially if they've just watched a film like 'It's a Wonderful Life' again). But many of us fail to keep that promise as everyday necessities and distractions get in the way, and from safety, an understanding of ABC (or short-term temptation) analysis, as above, explains how this happens really clearly. Forewarned about temptation theory can be forearmed but more practically, courses on time management can be life transforming.

Finally, it's worth stating the truth that for the vast majority of people, you cannot maximise your wellbeing alone. Try this simple exercise, which seeks to illustrate why we need to keep making the effort, no matter how difficult it is and how annoying other people are on any given day. Quickly write down your three or four favourite memories ever. The events that we hope we'll be thinking of on our deathbeds.

For many readers, every single answer will contain the word 'with.'

PHYSICAL AND EMOTIONAL WELLBEING

In the world of safety, disabled presenters such as Jason Anker MBE, who was paralysed in a fall at work, and Ken Woodward, who was blinded at work, articulate the importance of maintaining physical wellbeing in the most profound way. The first thing we need to do at work is to avoid physical trauma. Many lives have been blighted by merely seeing, or being responsible for, a physical trauma, as the Outtakes film 'The Witness,' about a colleague of Ken Woodward, makes clear. (I'm preaching to the already converted if you're reading this, I know).

Jason recently contacted the supervisor who felt guilty he'd not sent him home when he was obviously not fit for work. They had a good talk on the phone but Jason's offer of meeting up was rebuffed. The man (whose marriage broke up immediately after the accident), said he was 'just not ready yet.' The accident was 28 years ago.

COMMUNITY WELLBEING (OR 'GIVING')

This fifth element stresses that study after study shows that we need to *contribute*. Teachers, nurses, and safety professionals get an element of this from a job that is vocational, but many will get balance from voluntary work with charities and youth clubs. Or just supporting their family – especially if any specifically difficult needs apply. (It's also worth stressing that being actively involved in a safety process that helps keep colleagues safe also helps fill this need).

The nice lady who just this morning sold the office some shutters did a professional enough job, but only really came alive when she told the author about the mentoring work she does in prisons. More and more organisations now allow workers a period of paid leave during which they can work with voluntary service organisations (VSOs).

Indeed, as Generation Y come through, it's clear that many high-fliers who can choose are now actively using the quality of these schemes to discriminate between potential employers ... and employers are increasingly aware of such issues. I worked with an organisation once where the board signed off on a project heavy on soft skills and mental health. 'That's the CSR section of the company report I'm responsible for covered,' announced the CFO. I hope he was just articulating an appreciation of a win-win but, ideally, organisations will be a little less cynical in this arena!

A final point worth making. We'll stress later how important it is to ask for help when you need it and that setting up a culture where people are unafraid to ask is key. It really is often a win-win as being asked for help can satisfy this element of wellbeing.

INTERCONNECTEDNESS OF THE FIVE FACTORS

A recent study by MIND found that of the one in ten employees currently struggling with mental health issues, around a quarter said that it was because of problems due to work and half because of a combination of problems at work and outside. We'll be returning to this but it's worth considering some rather more practical examples here.

A study at King's College London found that when people sleep poorly, it affects the body's ability to regulate the production of hormones such as ghrelin and leptin, which control the feelings of hunger and fullness, respectively. Not surprisingly, they found poor sleepers consume an average of 385 extra calories the following day. It's worth considering the effects of an extra 35,000 calories on the ideal 'beach body' in a three-month period leading up to a holiday! Over a two-year period, that's more than 280,000 calories.

Another example: If you suffer from a bad back but hugely enjoy your work, then it's likely your colleagues don't even know about your back problems unless they find you doing stretching exercises in the corridor. However, if you don't enjoy your work, then it's likely to be a very different matter. (Indeed, Dame Carol Black has suggested the majority of people off work with 'bad backs' are actually struggling with mental health issues).

What study after study shows is that, whether causational, correlational, or a bit of both, everything seems to *cluster*. For example, people who exercise regularly are around three times less likely to show symptoms of depression. It's not as simple as 'everything is going well' or 'everything is going badly,' but there is a huge link. We're setting up and in the midst of virtuous and/or vicious circles every day. One of the practical problems this presents is that if you open up a gym on-site, you may be pleased to see that it quickly fills up – but it'll probably be full of staff already healthy and thriving, thank you very much. You'll probably find those in most need of exercise in the canteen or the smoke shack.

From HSE safety research, we know that a 'just culture' approach to incidents correlates with 'above the line' or 'discretionary' behaviour, and a decrease in such things as turnover, absenteeism, 'presenteeism' and spurious insurance claims. Perhaps the best-known model of a strong safety culture is the 'Bradley Curve,' with its concept of interdependency, trust, and 'brother's keeper' behaviour. We know that brother's keeper behaviours tend to occur naturally in a strong caring culture where they are modelled and facilitated – not in one where they are demanded.

This relates to the many wellbeing experts who consider the concept of 'care' as vital. As above, this book argues that an integrated, holistic, and humane approach to individual worker happiness is entirely congruent with low incidents and sustainable success generally. Specifically, that the elements of a thriving safety culture – learning, engagement, empowerment, and management integrity contribute to, and indeed are enhanced by, wellbeing directly.

POSITIVE PSYCHOLOGY: SELIGMAN'S PERMA MODEL

Another well-known holistic approach to wellbeing is the PERMA model, as described in Martin Seligman's book 'Flourish.' It comes at the issue from a slightly less practical viewpoint but overlaps hugely. PERMA stands for:

- Positive emotions.
- Engagement (i.e., 'flow,' where time passes quickly).

- Relationships (family, friends, or colleagues).
- Meaning.
- Achievement or accomplishments.

This short section is a consideration of 'meaning' which, although covered above already to an extent – not directly.

For example, a hard-bitten businessman or woman could lose himself in the act of the deal (engagement) and might derive a great deal of 'flow' and satisfaction from the achievement of stacking up a huge amount of money for its own sake. They may be 'content' but may well find rewarding relationships in short supply. Similarly, someone might be positive in the extreme, but with little meaning in their life. An aimless 'happy hippy,' if you would, although we're sure they'd point out their contentment is meaning enough, thank you.

What studies show is that engagement with work gives meaning, and that correlates well with productivity, but that engagement and positivity correlate twice as well with productivity. This means, ideally, with our eye on the bottom line, we need to address the job (it's meaningful and interesting), the person (who feels resilient and positive), and the culture (so does everyone else around me).

In short, 'we pay you well so what's your problem?' just doesn't cut it.

Taking a more personal approach, the famous Austrian psychotherapist Viktor Frankl, author of the multi-million-selling 'The Meaning of Life,' states that he found the will to survive Auschwitz so that he could rewrite a book just completed before he was interned and the book taken from him and burned on arrival. He did survive, and his writings have enriched the lives of millions – just as he hoped they would. A story that has always resonated with me and which sums up Frankl's approach to meaning:

Frankl had a patient who, after 40 years of marriage, saw no point in carrying on after the death of his wife. Frankl asked if the man would have preferred to die first to be spared the pain he was experiencing. Has was told, 'Of course not', he'd never want his much-loved wife to be going through what he was. Frankl suggested that this was the man's last act of love for his wife – to be the one who bore the burden of loss.

While we're in such deep waters, an interesting statistic: Childless adults tend to report themselves to be 'happier' than parents. They have more money, more freedom of choice, fewer responsibilities and constraints. Less worry generally. (Even more sleep, unless they take up the option to carry on partying!) But very few parents would even dream of swapping with them.

There is always a yin and yang and trying to summarise 'wellbeing' into a simple five-factor model is utterly impossible. It's too complex, nuanced, bittersweet, frustrating, contradictory, compromised, dynamic, and so on. However, the one suggested here: Money (basic needs), health, good relations with others, job satisfaction and giving, overlaid/interlinked with the more esoteric concepts of positivity and meaning, is, we hope, something we can use as a basic framework for a holistic approach here.

And again: When we're talking about influencing behaviour, the more holistic, integrated, and humanistic the approach the better!

PSYCHOLOGICAL SAFETY

A large number of bestselling books have been written on the subject (see Amy Edmondson), and some are considered controversial in the media. Viewed from the standpoint of proven safety excellence, however, I'd like to suggest that although the issue is indeed vitally important, it's not controversial at all really.

Fairness

Psychologists understand that a perception of fairness is perhaps the single most important psychological construct of them all because its flip side – blame and unfairness – has huge importance and reach. In short, back in cave dwelling days, if I, as a hunter gatherer ate everything I killed and didn't bring any back to those keeping the fires going, then they starve. If I return exhausted but empty handed and they've eaten all the stores and saved none for me … then I starve and they need to learn to hunt pronto!

So, co-operative species need trust, and most laws are basically an attempt to make sure we treat each other fairly. (Sometimes clumsily and sometimes with the powers that be especially well protected, of course, but that's another book!). What's key, though is that studies show we judge unfairness much more harshly than we judge illegality.

Blame

Elsewhere, we discuss the concept of 'Just Culture' and how studies suggest that 90% of errors tend to be caused by the environment rather than individual failing. We need to keep in mind the fundamental attribution error, which says we have a built-in knee-jerk response to errors in others of automatically blaming the person. (Especially if it's expensive, painful, or inconvenient). In short, the possibility of blame that is actually unfair is substantial. This is vital as nothing impacts on the unspoken psychological contract faster than a perception of being treated unfairly – or even seeing 'unfairness' meted out to others.

We can cross reference here to Freudian psychology relating to distortion, denial, projection, and the like which means that perception is seldom objective! However, 'Just Culture' based decision trees help to make the process of objectively responding to errors and incidents fairer, more transparent, and more consistent.

Therefore, I'd like to strongly suggest that any holistic wellbeing approach must have a Just Culture based decision tree as a key element (Figure 3.2).

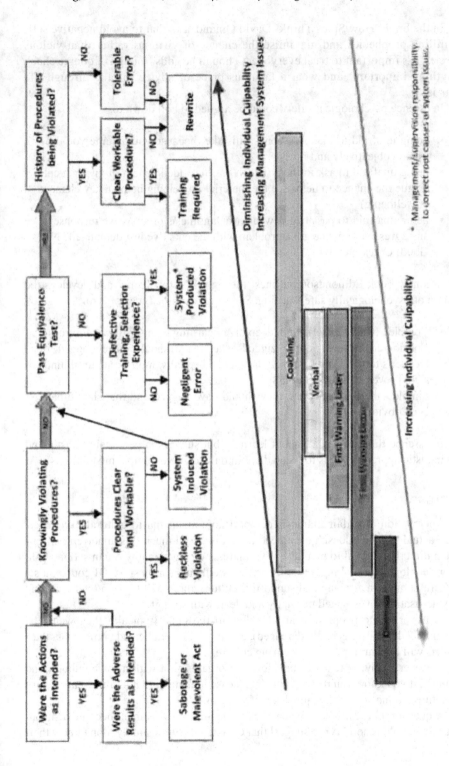

FIGURE 3.2 Decision tree model here. (See page 16 Definitive Guide Original)

In the book 'How Spies Think' David Omand says that to avoid negative self-fulfilling prophecies and maximise the chance of virtuous rather than vicious circles, it's important to trust everyone. (Though he adds 'though of course check anything important' and what a fabulous piece of advice to take through life this is).

In essence psychological safety could be argued to boil down to:

- being unafraid to make an honest mistake (because you assume you'll be treated objectively and fairly
- being unafraid to ask for help (which has the added benefit of giving a colleague the chance to help. See the material on helping being a key element of wellbeing)
- being unafraid to pass bad news up (the antithesis of a passive response to an aggressive "anyone got a problem with the cuts I've just demanded? No? Good! Crack on then ..."

In a new book Edmondson outlines 'the right kind of wrong' and develops the notion of psychologically safe learning. She suggests three types of error:

- Complex (where experts and 'deep dive' learning teams will be required)
- Considered (where people zigged when they should have zagged in Dekker's terms. Here both will have looked equally appropriate until hindsight proves otherwise). Finally
- Simple - they zigged when they should have zagged and that should have been obvious.

In essence this book is all about learning but suggests a genuinely holistic and humanistic approach needs to 'deep dive' into the cause of simple mistakes

BULLYING

If a perception of unfair blame is demonstrably anathema to cultural excellence, then actual bullying doesn't even need to be discussed much. It's abhorrent, is very often directly harmful to mental health, and has no place in any organisation at all, ever, let alone one seeking a high degree of wellbeing for its staff. Of course, it's a cultural issue – in a strong and supportive culture, there will be zero tolerance, and any accusation investigated promptly and dealt with strongly.

However, some people are just horrible to the bone (typically they would be accurately described as badly damaged), and it can't be eradicated entirely, but their efforts will be a minor issue in a caring culture.

However, bullying tends to be rife in an organisation with a toxic culture, and frankly, if you know you're in a toxic one, please remember that it takes years to turn a culture around so, frankly, just leave if you can.

A question in the media is when does demanding management become demeaning? Some 'old school' types suggest that Gen X or Z types simply 'don't know their born' and confuse the two … but I'll swerve that generational debate by pointing at basic 'emotional intelligence' skills. These are described later in the book, and I'd like to stress that the more of them deployed, the less need there will be to address who's wrong and who's right.

Of course, any companies HR department is likely to have extensive and detailed policies on such matters, and I think the only thing safety excellence has to add to the debate is an understanding of nuance and culture.

There is a scene in the classic film ('Happiness') that illustrates the nuance of bullying very well. In it, a table full of sisters are laughing, and the one sister not laughing is reassured 'we're not laughing *at* you we're laughing *with* you.' She responds 'but I'm *not* laughing.' This scene, for me, clarifies what bullying is and what it isn't. If the sisters stop and say 'sorry…', are mindful of her obvious discomfort and keen to make her comfortable again, it's just banter/humour that didn't entirely work. If they respond by ignoring her concerns and announcing 'then we're just laughing *near* you' (as in the film), it's certainly inappropriately dismissive at least, and it may well be bullying.

From a psychological perspective, an interesting approach taken by such progressive prisons is to assume, as alluded to above, that any bully is themselves in need of counselling. First, because coaching is always far more effective than criticism in changing mindsets, habits, and behaviours. (Note though this used *as well as censure* rather than instead of it).

The second reason is that most bullies were themselves bullied, so they are just acting out and passing on their own damage, and the sooner the chain is broken, the better.

(Though this is a safety book predominantly, I'd like to stress that the more hand in hand safety works with HR looking at genuinely holistic wellness, the better, so not something we can just 'leave to HR').

TRIGGERING

Any debate about what is bullying may well bang up against the current debate about 'cancel culture' and/or 'wokeness.' It's a hugely controversial subject, and only a full-on contrarian would be comfortable writing about it in 2023. Just last week, I gave the opening keynote at an event where the (excellent) closing keynote talking about leadership got no questions about his talk on the 'three key elements of leadership excellence.' Before anyone could ask him about it, a massive argument blew up about the fact that 80% of the speakers had been men. (Interestingly, an argument exclusively between women in the hall).

(There's an argument about the immediate benefit of putting up the 8 best speakers they could find and the long-term importance of quotas that I'll leave to the organising committee and entirely swerve if that's OK?!).

However, I'd like to argue that how to see the wood from the trees is something safety culture excellence is able to help with. Again, the basic premise of Just Culture is to see that context is king and that objective analysis paramount. As Sidney Dekker says – although with the benefit of hind-sight we can see a person zigged when they should have zagged – if we put ourselves in their shoes at the time, then both zigging and zagging would have looked equally appropriate. I'd argue that it's not the opinion but the value base from which the comment/incident occurred from that should be most important.

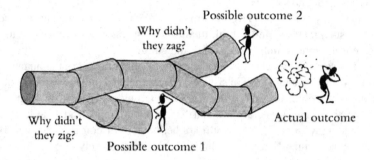

Possible outcome 2

Why didn't they zag?

Why didn't they zig?

Possible outcome 1

Actual outcome

FIGURE 3.3   Zig Zag illustration here. (See p. 59 Definitive Guide)

Political correctness is often described as 'gone mad' but anyone seeking a strong culture with staff wellbeing at its core should perhaps recall that nearly every progressive piece of legislation in history (in work or outside) was described at the time it was enacted by many powerful and influential people as 'madness.' 'Give slaves their freedom? Give 12-year-old chimney sweeps a day off a week? Let women vote? Are you mad? I mean I see your point to an extent, but if we agree to this change *where will it all end*?' In short, history almost always proves that the progressives were right, and it seems quite reasonable and logical, therefore, to assume that the progressives of the moment will be largely proved right too, regardless of how inconvenient their suggestions and demands are today.

But. That doesn't mean that all progressives are right all the time, and where they are wrong (I'd like to suggest) is where context is dismissed and where an objective analysis of what was said and *why* it was said isn't undertaken. Because it's clear that sometimes context *is* dismissed and knee-jerk blame reigns. (I'll *not* risk giving an example!).

The key issues should be: Were they trying to offend or were they stating an honest opinion based on a sound value base – but offence was taken? It's not possible to have genuine psychological safety and learning in an organisation where some are afraid to voice an opinion honestly and who genuinely intended to be constructive.

In short, if someone says something that some people find hugely offensive and complaints are made, then the process, as with safety failings, should be objective, contextual, just, and fair.

4 The Business Case

In these days of omnipotent social media a business really needs to ensure, as mentioned above briefly, that it addresses its CSR obligations well enough to avoid bad publicity and damage to the brand. Schneiders famous ASA theory says that certain people will be attracted to an organisation, that organisation will be minded to select those they feel will best 'fit in' and will in time tend to reject, or be rejected by, those that don't.

Oil and Gas and Renewable companies currently competing for the best young graduate engineers would be an excellent illustration.

Rather than just damage limitation, however, there seems to be increasing agreement that a 'happy' employee is around 10–12% more productive than the average worker, with an 'unhappy' worker around 10% less effective – giving a spread of around 22%.

Soma Analytics found in 2017 that UK FTSE100 companies that prioritise engagement and wellbeing outperform the rest of the FTSE by 10%, and a 2016 CIPD study suggests that stress and mental health cost UK businesses around £35B a year.

In short, the bottom-line figures are huge.

ABSENTEEISM AND 'PRESENTEEISM'

Studies suggest that for every day lost to absenteeism, around nine to ten are lost to 'presenteeism.' Turning up for work when ill enough to stay home may feel like a sign of commitment but may just be driven by fear. Certainly, all experts agree that employees turning up for work, but who are unproductive is a clear warning that something is wrong with that culture. This list of issues that flow from wellbeing is a bit like the scene in Monty Python's 'Life of Brian' 'What have the Romans ever done for us?' Engaged and motivated workers:

- Work a little harder; and
- Leave less frequently (and it's always the best and most attractive to other organisations workers who go first).

But they also:

- (Need to) take less time off sick;
- (Genuinely) recover from illness and injury more quickly;
- Prove better trainees; and
- Contribute better to team problem-solving.

and

- Apply more discretionary effort generally.

DOI: 10.1201/9781003449997-6

What Has This Wellbeing Ever Done For Us?

"OK, fewer of our best staff leaving. A reduction in absenteeism and presenteeism. More creativity, discretionary effort, productivity and a better reputation ... granted ... but apart from that ..."

FIGURE 4.1

In short, they are pretty much better at everything. They are even far less likely to have a second heart attack and die after a first attack. As alluded to above, it's been estimated by the health guru Dame Carol Black that for every ten workers off with a 'bad back,' eight are off because they're somewhere between 'fed up' and clinically depressed. Linked to this, and unsurprisingly, positive people succumb less often to mental health problems.

Demonstrating the sort of savings involved, Seven Trent demonstrated that a 1% reduction in sickness absence from 4.3 to 3.3 meant a £1 million a year saving.

The Harvard Business Review describes successful wellbeing (WB) programmes as being aligned with overall company identity and goals, having engaged leadership, being comprehensive in scope and quality, and easily accessible in terms of access and cost. As well as agreeing with the above lists (indeed, largely *because* of the above lists), the Harvard Business Review study suggests a return on investment of '300% or more' and companies on the Harvard Business Review 'Best Places to Work' list outperformed the market average by 115%.

In the UK and elsewhere during the COVID epidemic, a lack of discretionary effort and engagement had people moving from pub and café to pub and café, signing in their details accurately as they went, sure that sooner or later they'd get a 'ping' that meant ten days isolation (or 'isolation') required. (Motivated and self-employed people generally wrote the wrong number down!) As things returned to normal, the 'great resignation' began, and it seems more than half a million people in the UK have effectively said, 'that's enough of that!'.

In short, wellbeing is not a cost it is an investment.

A KEY LESSON FROM SAFETY

We've often talked about *halving* a problem (usually lost time accidents, sometimes fatalities, and sometimes near misses) as a 'step change.' Many, many companies in the UK and around the world have achieved this in a year to 18 months, then repeated that success. Indeed, several industry-wide initiatives have been labelled 'step change' with the same aim and have succeeded. Halving a problem is not unrealistic pie in the sky – it's achievable. Keeping this in mind is good, as simply halving the gap in engagement between the UK and the G7 average would be worth around £200 billion to 'UK plc!'

To reiterate how achievable that is, we're not talking about excellence and best practice here; this is just for getting to a place on a par with average countries or only being half as bad as the UK is now! It's a huge 'war chest' to play with and not difficult to sell it as an 'investment.' Indeed, the world of H&S has never been more bullish when given the chance to look a CFO in the eye and make a pitch.

CASE STUDIES

A UK builder described at a conference how their industry was bedevilled by a high churn of staff because an offer of £2 an hour more was enough to induce a worker to move on. After the introduction of a systemic empowerment and enrichment approach, turnover decreased significantly, because now an extra £2ph *wasn't* enough to induce an electrician or a bricklayer to walk away from a job they now enjoyed. (Key here, it was suggested, were the installation at workers' camps of soundproof rooms with excellent Skype connectivity that could be booked in 20-minute blocks by migrant workers for conversations with family at home. Related: In 2022 all ships around the world do not have Wi-Fi. In the maritime industry – where wellbeing is a big issue as around 1 in 4 are thought to be 'struggling' – it's well-known that these also tend *not* to have anywhere near the best and most sought-after staff).

Another example. The UK care industry is in utter crisis. Private investors are leaving the market in droves, and public providers are withdrawing services across the board. Around 900 staff are leaving the sector every day. Bucking this trend by taking a staff wellbeing-centred approach to sustainability, Sandwell Community Caring Trust has two key indicators: Staff turnover and sickness.

As well as training, staff are paid as much as is viable, given extra holidays and benefits, and are consulted and involved in the governance of the business. The reduction of turnover from 40% to 10% means recruiting 200 fewer staff a year, and reducing sickness decreases the reliance on (more expensive) agency staff. These time and cost savings are focused on wages, training, and asking the cared for what they need.

The CEO, Geoff Walker, says that it makes a low-margin business sustainable – or, the impossible just about viable. The focus has to be on the quality of care, and where 15-minute social care visits from badly motivated and ever-changing staff is increasingly the norm, it enables perhaps the most important key performance indicator (KPI) of them all: An increase in the quality of interactions.

(In the care industry there's a well-known expression 'every contact counts!'. We've started using it on *all* culture creation courses).

PROBLEMS ORGANISATIONS FACE IMPLEMENTING WB PROGRAMMES

The question is begged: Given the figures above, why isn't *everyone* doing it, and doing it well? Part of the reason, of course, is that it isn't as immediate as safety, so that it's difficult to see the damage that's being done and the opportunities lost. And, of course, with home life and work life evermore blurred, especially following the COVID surge in working from home, the demarcation is even less clear than it was, so it's hard to say for certain what's causing what, and, therefore, who's responsible legally and/or morally?

The well documented and researched experience of **occupational health** shows how this works day-to-day. In the UK last year, a conservative estimate (Rushton Report) is that around 13,000 people died of occupational exposure, and typically from exposure decades ago. This is often exacerbated by smoking, drinking, poor diet, air pollution, and stress. Though asbestos and other issues are now well-known and controlled, many new risks are entering the market with the advent of nano-technology, and the experts suggest that the 13,000 figure is likely to increase, not decrease. The problem is that the politicians and CEOs in charge today will long since be retired when the price needs to be paid.

We have come a long, long way addressing the S in SHE, but there is still a long way to go towards the H.

Specific problems regarding the effective cooperation in the design and delivery of a co-ordinated and holistic approach to the wellbeing of staff, articulated by the Campbell Institute and others, include:

- It (typically) needs HR and SHE to cooperate, or at least for HR to have the practical field experience a good-quality SHE department will often have.
- It can run headlong into union suspicion (and/or a resource/political 'turf war').
- The people it most needs to target (negative, middle-aged men) are the people hardest to reach.
- Tracking progress can fall foul of data privacy issues.

One example of how not to coordinate HR and SHE is the major European airport that rolled out some behavioural standards through HR that included reference to six core values, including 'safety,' but without consulting the SHE team in any way. The safety department therefore struggled to support their HR colleagues not just because of legitimate 'buy-in' issues, but because the definition of said safety behaviours was, to quote the head of safety, 'utter crap.'

The writer Gareth Morgan ('Images of Organisation') sums it up well when he suggests that resources and decision-making is power, and resources are always scarce, so people and departments will be inevitably disposed to compete for them. In short, politics is inevitable but not addressing this pro-actively and, in an adult, co-ordinated fashion isn't and simply shouldn't be allowed. At a strategic level, organsations are set up to address their risk register the best it can. Who does what and under what banner should be irrelevant, shouldn't it? More than that, we know that 80% or more of training efficacy is in the follow up and embedding, and with W,H&S training becoming ever more generic, that really requires a co-ordinated approach that includes the appraisal and personal development systems. This means HR and W,H&S must work together.

Also worth noting is that in any decent culture, the SHE team will have an undiluted 'here to help' credibility that a multifunctional HR team might struggle to match because the HR team will have to lead discipline and the communication of bad news. (Though in a really poor culture, the SHE team may be seen as an 'elf and safety' inconvenience, of course).

Politics may also apply to unions too, who may be wary of management initiatives – not least because of the hard-won experience that suggests they *should* be wary as not all management are benign of course! On the other side of the fence: I once had a senior union official say to me at a nuclear site without even a hint of irony: 'If there's any empowerment work to be done here with my members, I'll be the one to decide who, when, and how.'

The solution, as ever, to turf wars and other politics, is strong executive leadership – and this is one of the key reasons why executive buy-in to a holistic plan is so important. It's just a variation on one way Daniel Kahneman describes what he means by 'slow' executive thinking. He suggests if a major change will work well for seven out of ten companies in a group, but fail for three, then no one individual company will ever choose to take a risk in the face of such two to one odds. This needs to be an executive decision, properly co-ordinated, supported, and the roll-out learnt from.

The same is true with wellbeing and, as above, studies show clearly that where it is part of the strategic executive level planning process, those ROR figures hold true. Less true where it's addressed only at a tactical 'bikes and bananas' initiative level.

DEVELOPING A WELLBEING PROGRAMME

It was tempting to add a subchapter here going back to very basics by walking through a basic process implementation approach. But, in truth, there are 1,001 variations on the: Plan, do, monitor, review process available in 1,001 places if you should need them. You will, of course, need to plan, prioritise, resource, monitor, and communicate effectively, etc. as with any process. However, there's a lesson from safety excellence that is hugely useful.

This is to use basic risk assessment techniques in conjunction with a bow tie approach. A simple example from the front line of stress would be assessors reviewing potentially unacceptable online material – often alone in their own homes. Clearly, trauma, both instant and cumulative, is a possibility, and many companies have switched from showing the material with the sound switched *on* as a default to the sound switched *off* as a default (as often there is no need to hear the material – viewing it is sufficient without the added trauma of hearing it too).

So, as well as trying to minimise exposure before the event (the left-hand side of the bow tie), such as pro-active employee assistance programmes after the event, represent attempts to minimise any harm.

A good bow tie shows, in a user-friendly way, exactly what we're trying to achieve and who is responsible for achieving it. You'd never operate such an oil rig without one – but, as above, when it comes to accidents and illness, process safety will always be central to any risk matrix … but in terms of harm to people, it's probably not as important statistically as many other factors.

5 'Big Blue Pie Slice' Days
The Individual in Wellbeing

MISTAKES AND ERRORS – THE BASICS

The human brain is a wonderful thing – the cleverest thing in the universe by a country mile. However, it can be put off, backfire, malfunction, and tire easily. It's also really easy to distract, even in safety-critical situations (as the 'hello boys' advert of a few years back demonstrated so amusingly).

In short, on a typical day, studies suggest we'll be 'away with the fairies' or in 'zombie mode,' on average, for five to ten minutes each hour. This is bad enough for a single team of employees in a van, but for organisations that have 10,000 employees worldwide, that's an awful lot of risk mounting up day-by-day.

Often, a company will notice that the accident book contains nothing but 'silly, easily avoidable' accidents, and some form of exhortation to 'just please pay full attention and take care' will be made. The trouble is that there is simply a physiology-based upper limit to the effectiveness of this tactic, and as the world of work becomes more mechanised, error management has become more about 'mistakes of thinking.' The same principles apply.

In safety, we've long known that rather than say 'pay full attention at all times' it is far more effective to say, 'if you see a trip hazard when you're bright and alert, stop and clear it up, so that it's not there to trip over when you come back around the corner ten minutes from now "away with the fairies"...' The same principle applies to such as communication, dialogue, decision-making, and project and change management. Minimising cock-ups needs good habits and tactics dictated by a clear appropriate strategy.

THE PERFECT WORLD

This limitation still applies to people who are 'still you ... but on a good day' as 1,001 vitamin and supplement adverts have it. You are happy in your relationship, your family is well, *you're* healthy, you're not on medication, you have no money worries, you're fully trained, good at your job, and you enjoy it. You have a suitable level of autonomy (not too little, not too much), and you know where that job fits into the bigger picture. In short, you're not so flat out all the time that you can't even catch your breath, let alone *reflect*.

Most importantly, your direct boss is emotionally intelligent. A pro-active and skilled communicator who gives negative feedback when required without personalising, generalising (or breaking any other golden rules) but who is always looking for a chance to give positive feedback. Calm and assertive and always 'adult-to-adult' they are a skilled coach who asks questions to draw out knowledge and to deliver

DOI: 10.1201/9781003449997-7

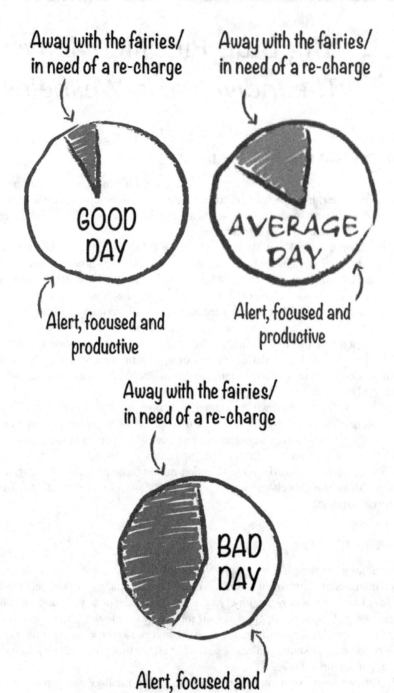

Away with the fairies/
in need of a re-charge

Away with the fairies/
in need of a re-charge

GOOD
DAY

AVERAGE
DAY

Alert, focused and
productive

Alert, focused and
productive

Away with the fairies/
in need of a re-charge

BAD
DAY

Alert, focused and
productive

FIGURE 5.1

'discovered learning' moments. They always look to build in leeway and pro-actively asks, 'what's key to a good day for you?' or 'what do you need from me so that you can work efficiently and safely?'

When something goes wrong, they always first ask 'why?' *curiously* (not jumping to blame but instead have the default of assuming there's a contextual reason). They pro-actively ask: 'Anything slow, inconvenient, or uncomfortable about doing this job safely?' knowing that if there is then it's just a head count of how many people are cutting corners.

In short, you work in an empowering, supportive culture – not a blame ridden toxic one.

BAD DAYS

At an individual level, we can talk about working for a boss that shouts a lot and is a low-grade bully. Or you may just be hungover from celebrating. Or hungover from drowning your sorrows and/or because you have a drinking problem. Or you may have relationship or financial worries. Or you may be on long-term medication for chronic illness or pain.

The list is, of course, endless and individual, and so should be the responses. Ideally, as well as tailored to the person, these responses will ideally be as caring, objective, fair, consistent, and pro-active as viable. However, there are several issues that are more systemic and predictable, and so responses should *definitely* be systemic and pro-active. They are all predictable risks, and the world of safety has long known exactly how to deal with predictable risk!

The four main causes we can anticipate pro-actively and address would be issues from outside work and/or fatigue and/or working in a toxic or just unsupportive culture and/or mental health. It's the interplay of the four – and what organisations can do about it – that this section addresses.

Some more specific issues that we've begun to address more often recently:

The Menopause and Menstruation

In recent years, we have finally started to talk about the menopause. In the UK, some 4.4 million women in work are of menopausal age, and 10% of women have left the workforce because of it. The cost to 'UK plc' is estimated to be around £1.9 billion annually. There are confidential surveys, apps, and courses on how to talk about it, but not all organisations are using them …

Did you know that Russia introduced a policy of menstrual leave in 1922 and Japan in 1947? It's still rare, and controversial, in the western world, however, and studies suggest a majority of women remain uncomfortable talking to their manager about it. Again, the costs are colossal and, again, ignoring it and crossing fingers is the *least* cost-effective way of dealing with it.

Neuro-Diversity

Most people are neurotypical, meaning their brain functions in the way that society expects. One in seven people are neurodivergent, however, meaning their brain functions differently in terms of learning or processing information than is considered standard or typical. Typically, the list includes dyslexia, autism, dyspraxia, and ADHD.

A key theme you might have read about is to view ND people as having 'super-powers' that can help an organisation thrive. Dyslexics tend to be creative and good at people skills – Richard Branson being a famous example. Autistic people can often be very reliable, detail aware, and have a laser focus that allows them to develop great expertise in a field. (Alan Turing, of the film, 'The Imitation Game' and who essentially invented the computer, being a famous example). And you really don't want to be in the way of someone with ADHD (Attention Deficit Hyperactivity Disorder) when they get the bit between their teeth as they add drive and resilience to laser focus. Most organisations could find a use – *in the right role/department!* – for the likes of John Lennon, Michel Phelps, Thomas Edison, Albert Einstein, Salvador Dali, and Winston Churchill.

The italics are key. If we simply expect ND people to be NT when we want them to be, it'll cause problematic issues, if not chaos. So, we must, as part of our holistic plan, know that, for example, promoting a superb but ND engineer to manager (especially if we don't train them in people skills) will almost certainly end badly.

Or, we might need time management training (and/or software) for those with ADHD and noise cancelling headphones for anyone with OCD. We might want some people to write talks and others to deliver them – some to prepare materials and agendas for meetings and others to chair them.

A simple example. I worked with a company once where a chap on the autism spectrum and with learning difficulties was considered their best ever table clearer. (More than that, happy to do this job indefinitely). Then they took away dedicated lockers, so he turned up earlier and earlier so he could use 'his' and when he couldn't, he got very upset and agitated. It was only weeks into his unexpected plummet in performance that a manager noticing him arrive *hours* early and thought to ask why.

Here's another example: The actor and patron of OCD-UK, Ian Puleston-Davies, used to play the builder Owen Armstrong in Coronation Street (the world's longest running TV soap drama). If you're a regular watcher, you'll have seen him standing at the bar, supping a pint, and arguing in the Rovers Return pub a thousand times. Or standing in the door of the kitchen in his house holding a 'cuppa.' But *always* standing as he has a quite severe OCD about sitting. Two observations if you were a watcher. First, there's a very good chance you thought Owen was a great character that added greatly to the drama, but secondly, in all the years he was in the pro-gramme, you never even noticed he was standing in every single scene!

Obviously, this is a huge and complex field, and there's any amount of excellent guidance on the shelves of the HR department. But from safety – we know well to ask 'what do you need to thrive?' (After all, nearly everyone will have something they need). With that mindset to the fore, I'd argue that ND issues are just a subset.

MENTAL HEALTH

DEFINITIONS AND INCIDENCE

Having a bad day isn't good for output, relationships, or job satisfaction. But at an individual level, the company's share price isn't too much of an issue to you personally if you're moving from 'bored, frustrated, and fed up' to 'struggling' with mental health or worse.

How many colleagues are 'struggling' with their mental health? (A term that broadly means MH issues are getting in the way of you functioning relatively efficiently). It depends on how you define it, of course, but the charity Mind says one in six in any given week and one in four in a year. Whatever the actual ratio, in Europe alone that certainly accounts for tens of millions of people on any given day. At the very sharp end, more than 4,000 working-age people in England and Wales took their lives in 2019. Compared with current workplace fatalities in Great Britain (142 in 2019), that's a ratio of more than 30 to 1, and unlike long-term health issues (as discussed above), that *is* an immediate problem for companies. (These figures skew towards middle-aged men and people in poorly paid work).

Risk factors include sensationalist media coverage, childhood trauma, issues regarding the misuse of drugs and alcohol where 'self-medication' may be involved, relationship problems, and of course 'basic' mental health – however we might define that.

Again, with these figures in mind, I'd like to again stress that people who have a job they enjoy tend to be mentally healthier than those who don't need to work at all. The reverse is true of those who dislike their work.

Just this morning, I had a client say that a friend of his killed himself over the weekend by throwing himself in front of a speeding train. This friend was 'always upbeat and positive and the very last person I know I'd have thought might do something like this.' He leaves a wife and two children behind him.

And it's getting worse. In the UK and around the world, the side-effects of the pandemic and COVID (and long-COVID) impacting substantially on mental health. Before the pandemic, self-harm in young adults was at record levels in the UK, as were eating disorders. A sobering statistic, in recent years, more UK students have taken their own lives than workers have been killed in the workplace, according to the UK Office for National Statistics (ONS). Young adult mental health services in the UK (CAHMS) were already officially in crisis before COVID-19, and the current energy/financial crisis is projected to last long after this book is published.

I'd like to argue that it's well worth remembering that these young people are the next wave of workers heading straight for our organisations so the message to organisations is: Mental health in the workplace is a huge problem, it's getting bigger, and you're probably going to get very little governmental help.

There was already a sea change underway, but COVID-19 has changed, perhaps forever, the way we consider mental health. As above, it was often assumed that most causes of mental health issues were physiological and/or related to factors outside of work, and simply no business of an employer. There was also a concern that suggesting a link between mental health and safety might lead to active discrimination, and therefore make employees wary of being open about any issues in the workplace. More than that, it was also argued that with life-changing accidents and fatalities plateauing, there was still plenty of pressing safety-based work still to do, even in the UK, a world leader in OSH.

However, many people spend more of their waking hours at work than they do at home with family and friends. By taking a holistic and integrated approach to human error generally, we really should treat day-to-day mental health issues as contributing to risk in the same way as fatigue, lack of resources, lack of training, and a poor safety culture may do.

It's mental health as well as safety and with the aim of impacting on safety – not *instead* of safety.

We simply need to reduce the risk by addressing it objectively, pro-actively, systemically, and with practical tools. That's just good risk management. The good news is that the world of safety, health, and environment already has proven experience of using many of these tools and vast experience of the classic pitfalls organisations make when seeking to address 'people' issues.

What Is Mental Health?

It's worth remembering that we all have mental health, just as we do physical health, and it is equally variable.

For example, do you ever suffer from any of the following?

- Memory lapses.
- Disorganisation.
- Over confidence.
- Feeling anti-social.
- Lack of appetite to groom and clean.
- Feeling agitated and irritable.
- Feeling lethargic and apathetic.

Personally, I score a weak 6 or 7 on all but my very best days! (And a stronger 7 on a bad day).

The reason for this little exercise is that if you add hallucinations and delusions to that list, that's all of the main symptoms of schizophrenia. This illustration isn't to minimise the seriousness of what can be a horrendous illness but to point out that we're *all of us* on a *continuum* and that tends to change week by week and day by day. For example, there's a good chance that there will be days when someone formally under treatment for schizophrenia will score better than you and I on every item on that list above. (Because they are taking their medication and are careful about sleep, diet, and exercise).

Likewise, it can be argued that depression is simply thinking overly often of things that have happened in a negative way and anxiety thinking overly often of things that have yet to happen in a negative way. (These such as regret, bitterness, and sadness. Or fear, anxiety, and stress). A cognitive behavioural therapist would stress that every minute spent thinking negatively of things that have passed robs us of a minute to enjoy the 'now.' (A brief and appropriate prompt to learning aside). Likewise, every minute spent thinking negatively of something yet to happen steals a minute from us working to make sure it doesn't. (Again, a brief and appropriate prompt to excellence aside). In both cases, once any brief prompts are over, we need to be in the moment. (Hence 'mindfulness' which you'll almost certainly have heard of!)

There is a huge range of factors that can impact positively, or negatively, on our mental wellbeing, and it can have just as profound an effect on our bodies as physical illness does. The difference is that there is a stigma attached to mental health, a social taboo that can make people ashamed to talk about it or seek help. This is compounded by the lack of obvious physical symptoms. If someone has a streaming cold or is limping when they walk, others will notice and (hopefully) offer sympathy or assistance. The trouble is that you can't always see when someone's mental health is poor, so a great many people are left to suffer in silence, and, as above, this can have tragic consequences if it isn't addressed.

BAD DAYS, HUMAN ERROR, AND INCIDENTS

As above, mentally struggling contributes to a bad day just as fatigue, lack of time and resources, home distractions, a toxic culture, and disempowerment generally do.

Overall, three behavioural issues seem to link mental health and accident and emergency admissions.

1. A lack of focus and situational awareness: More likely to be oblivious to risk.
2. An increased fatalism and disengagement: Less likely to care about risk.
3. An increase in poor decision-making and an increase in poor interactions. (i.e., the individual creates risk/misses an opportunity to manage it better).

It must be stressed that it's not as simple as saying depressed people are more likely to be fatalistic and anxious people more likely to be distracted, because anxious people can be hypervigilant (although sometimes about relatively trivial issues – distracting them from real risks). And many sad people know that helping others makes them feel better so are, actually *more* likely to be pro-social.

There's very little consensus in the data, but an Australian study of 60,000 workers by Hilton and Whiteford suggests that incidents increase by between 50% and 150% depending on the situation and the severity of the issues being faced. It also suggests that individuals are only 70% as likely to contribute 'successful' behaviours to the organisation when they are faced with mental health issues. Successful behaviours are classed as discretionary behaviours that contribute to positive 'culture creation,' such as creativity, looking out for new starts, challenging, or volunteering. Then there's the truism to always keep in mind that 'compliance itself is often discretionary.'

I suggest a very rough rule of thumb is that on balance 'struggling' workers are around twice as likely to have an accident and half as likely to contribute 'discretionary' effort in a positive way. (I say this knowing the common sense can be wrong and face valid isn't always reinforced with hard data, so I'm fascinated to see what coming research finds. For example: A meta study by Seligman found, to my surprise, that positive people have as many accidents as everyone else. It turns out their optimism can work against them and that 'getting stuck in' has a downside. Which actually makes sense when you think about it – as does the finding that they recover from any injury twice as quickly!)

So, I must stress again as strongly as I can that I'm *not asserting this to blame individuals* in any way but to encourage organisations to address this widespread risk issue pro-actively, objectively, and practically. It simply reduces risk much more effectively than finger crossing.

WHAT CAN ORGANISATIONS DO?

A wellbeing strategy must be something that is developed and reported at the board level as a key element of organisational sustainability and excellence. We don't want control based at a lower, tactical level, where we have reactive Employee Assistance programme hardly anyone contacts, and 'bikes and bananas' initiatives whose primary benefit seems to be free fruit and subsidised gym membership for those employees who already eat healthily and take exercise!

Ideally, this will be strongly branded and supported by personal testimony stories from senior management. Of course, there must be health screening, subsidised gym membership, free to attend financial wellbeing talks, meditation classes, bananas in reception, and bike riding schemes. You will, I'm sure, know the sort of thing!

Ideally, however, these facilitation of health approaches are because the workforce has suggested it would be useful rather than having it imposed on them.

The key thing here is not to think that this element is the most important as culture is always the most important element, but before addressing that, we need to talk about individual approaches.

INDIVIDUAL RESILIENCE

It's often said by wellbeing experts at conference events 'resilience for individuals last if at all' but many companies still address wellbeing through seeking to make the individual more stress proof.

Some of these courses are run by ex-SAS soldiers or Olympic athletes and are, of course, often hugely impressive and can indeed be life-changing for an *individual*. They focus on such as:

- Knowing yourself and your impact on others.
- Habits that give people health and energy.
- The power of a positive mindset and being in the moment (hence self-affirmations, meditation, mindfulness, and the like).

But it's often said that 'culture is king' and a key lesson from behavioural safety is that it's the environment, not the person, that mostly drives behaviour. (As above, many BBS approaches, for example, are rightly criticised by unions for focusing on the individual and looking as if they're 'blaming the victim').

Like companies that had an inspirational speaker give a powerful 'be careful' talk that covers safety while doing nothing to minimise the risks faced by employees at source, organisations that roll-out resilience training but do nothing to reduce stress at source are, I argue strongly, guilty of focusing on the wrong thing. They should first seek to improve the culture generally and to tweak the role to suit individuals. If not, they are, to use the technical psychological term, 'magic bullet merchants.' Or to quote a delegate I interviewed once:

> *The course was great but I'd only need half as much resilience if my boss wasn't such a total **&%$.*

Indeed!

A case study from a long time back featuring a company difficult to work for seeking a 'magic bullet.' I ran an individual resilience course and used the exercise 'write your own ideal obituary' (As in, 'if you're not heading towards it, you're probably feeling low in traction and meaning'). The star graduate on the course (a double first from Oxford and a sporting blue as well) came in on day two to say goodbye. The exercise has prompted her to have a good think overnight and she quit on the spot. (As you can imagine the client subsequently asked if I'd focus on breathing exercises and the like and cut out the more holistic pieces).

Appropriate individually focused elements that can be of use, of course, either in leadership training or for all employees – *but ideally, always as part of a holistic approach*. But there is one element when focusing on the individual in the short-term is entirely appropriate. That's if, for whatever reason, they are 'struggling.'

Addressing Mental Health – A Holistic Organisational Approach

As well as the strategic element discussed above, I'd like to suggest just two more basic strands that fall squarely under the culture banner.

The first is to pro-actively identify colleagues who are starting to struggle with their mental health or who are disengaging from the organisation generally. Mental health first aiders can help with this, but they are not a magic bullet, and having all leaders actively involved (as with a 'felt leadership' safety approach) is better.

The second strand is to systemically strive to create the generic benefits of 'good work is good for you' by developing a generic, strong, and caring culture based on objective analysis and empowering and engaging leadership.

It is this strand that seems to be missing in many company's wellbeing strategies but I'd like to argue that it's the very engine of the win-win I keep emphasising. Although many mainstream wellbeing strategies and approaches I have seen mention culture – they have all meant culture as in individual differences and mindsets. (Race, ethnicity, etc.) They don't mean culture as in the contextual 'culture is king.' This, I argue, is perhaps the most important cross-over lesson from proven safety excellence.

Talking to a Colleague

First, to address at a day-to-day individual level when a colleague is struggling.

What organisations such as the Samaritans stress is that often a simple conversation about how they feel is all that's needed to give them pause for thought. (The Samaritans in the UK alone have more than 5 million contacts a year).

However, all colleagues are very well placed to pro-actively notice there may be a problem and initiate what might be a life-saving conversation. The Samaritans also stress that mentioning suicide specifically is not the 'worst thing we can do' as many instinctively feel, but may open the door to a vital conversation. The worst response you'll get is 'no … not at all!' from someone a bit indignant at the suggestion. But the best you might achieve is that you might save a life. Indeed, it's suggested that the likelihood of saving a life by, for example, approaching someone looking poised to jump from a bridge far exceeds the likelihood of saving someone who has had a heart attack with a defibrillator, though nearly all readers would attempt the latter. (At a conference, I was told by a speaker from MIND that it's 50% versus 12% or so. That's circa four times more likely to save them. However, though I've struggled to verify that data, as the well-known safety principle always says, don't find a reason to walk past).

Experts stress that listening is all that's required. Specifically, we shouldn't reassure them 'it'll be ok,' as we can't know that. Nor should we even talk about personal experience. For example, 'I know just how you feel'...or 'you remind me of Uncle Arthur and he's fine now' ... this just makes it about you and sets them up to say 'thank you, I'm sure you're right, you're very kind ... thank you' without actually helping them at all. Instead, always talk about them. If you have personal experience, use that to fuel empathic listening.

We just need to ask open-ended questions about their feelings and *listen*. (Listening, as a generic skill will be addressed a little later).

'I'm Fine, Thank You': The Problem With Men

A basic 'walk-and-talk' tour of the shop floor will, 9 times out of 10, generate the answer 'I'm fine thank you, how are you?' to the question 'how are you?' This is especially true of men, who can be hugely reluctant to open up, even when in the depths of despair. The basic approach is to ask open, high-value questions, not closed, low-value ones. ('Are you OK?' is an example of a closed, low-value question as it has a yes/no answer. An example of an open, high-value question might involve asking someone what they think or how they feel.)

So, Ask 'How Are You?' *Twice*

In my book, *Talking Safety,* I suggest asking 'How are you?' as an ice-breaker before getting into analysis and coaching. Of course, the answer to this social nicety is nearly always, as above, 'I'm fine, thank you ...' For western men of a certain age in particular, it's virtually a law! However, once some rapport has been developed, it's worth asking it a second time with eye contact and empathy. You may again well get 'I'm fine' from someone who isn't fine at all, but this second request is much more *likely* to start a conversation that saves a life.

'FORD' is an excellent acronym reminder of how to instigate small talk. Simply asking about family, occupation, recreation (sports/hobbies), and dreams and ambitions. (Though in an already work-based dialogue that's FRD!)

Personal Testimony. (Another Lesson From Safety – Authenticity)

Approaching the target audience in the right way is key, because knowledge is only base one, and the people we most want to reach are also the ones most likely to dismiss out of hand anything not pitched in the right way. The three experienced workers that sit at the back of a briefing, arms folded, when being told about the risk of doing X and Y, who then go out and do X and Y, are the same who sit at the back taking in wellbeing advice with a shrug.

We quickly learnt in safety that if you want to reach people, you have to never patronise them and engage in dialogue in a language they understand. Talking to people like adults and using data and illustration helps maximise the chance of 'discovered learning' leading to a genuinely internalised promise and this is the only sort we ever really keep.

The safety world has long used accident victims to drive home the personal cost of an incident, and increasingly the wellbeing world is using the same approach. Ideally, this won't just be people recovering from a breakdown, but more pro-active testimony. The well-known actor Ian Puleston-Davies, mentioned above, gives humorous talks about how he copes with his OCD on film set. He is always inundated with fellow sufferers following a talk.

James Reason coined the term 'safety is a guerilla war.' In safety, we long ago learnt to use what works, but much wellbeing work is still very polite, formal, and restrained. We need to rediscover the 'at all costs' attitude of someone like the controversial and legendary safety campaigner James Tye (who founded the BSC) and get stuck in. (Perhaps his most famous stunt was to address the 'elephant in the room' by marching elephants down Whitehall to the UK parliament). A more practical example: One company I know of said they'd had a fantastic response to a prostate and testicular cancer awareness campaign. Rather than Occupational Health, they used a professional comedian, who effectively did it through the medium of several 'knob jokes.'

It was, I'm assured, a very successful approach!

The FI Toolbox Talk

Likewise, my consultancy's most successful methodology is probably the 'F I' toolbox talk, which was invented by accident in discussion with my partner Jason Anker MBE.

One day, Jason and I were discussing his accident and specifically, the fact that Jason had, as he has so often encouraged others to do, taken 5 seconds to consider the risk associated with an unsecured ladder on unstable ground – but then climbed it anyway (and fell and paralysed himself). His honest answer to the question 'why?' became our *Fatalism and Intolerance of Individual, Organisational, and Societal Stressors Index* or the 'F I' index for short.

Mostly, we were, of course, seeking a laugh as the audience got the joke as to what F I stands for – but a very interesting thing started to happen. Clients reported that they really liked our big blue pie model of level of individual focus, but they didn't ever run 'Blue Pie' toolbox talks. However, they did run 'F I' toolbox talks. After the fourth one to report that they worked really well and so we formally incorporated it into our human error training courses.

More than that, what we found was that human error leadership course after leadership course rated it as the top 'golden nugget' that they intended to use as soon as possible. (At the end of any course delegates are given free choice to commit up to three things they will do straight away).

When we thought about the social science behind this, we found that there's a bit of a link to such as neuro-linguistic programming and other interpersonal sciences. If I'm having a bad day I'm not sitting there thinking 'oh gosh, I'm really having a big blue pie slice day today' I'm thinking something much more base and earthy. I'm thinking 'f*** it.' Basically, if we talk to people using the words that are *actually in their heads*, we are much more likely to connect with them.

In addition, I once gave a paper with an award-winning comedian Steve Royale on the science behind humour, specifically covering how a shared laugh is the fastest way of communicating that we both 'get it.'

When we're sitting there struggling, feeling someone is 'getting' us can make all the difference in the world. We've long known that pro-active conversations about safety are always better than reactive ones, and the same is true of mental health and wellbeing. F I toolbox talks are just a user-friendly way of starting those conversations.

A Case Study in How *Not* to Do It

I once reviewed the approach of an organisation that was very proud to have produced an excellent wellbeing webinar that talked employees through key resilience techniques and lifestyle. They'd made a point of keeping it available indefinitely after the first broadcast so we watched it. It really was very good indeed! However, it was also optional. So, quite aside from the issues that 'education is but base one,' you won't be surprised to learn that it had been watched a mere 60 times to date. The company in question employs more than 10,000 staff, and we're pretty certain that few of the 60 were from the population they most needed to reach.

Such films are simply not enough. We need front-line supervisors running F I talks and colleagues in the habit of asking 'you got time for a coffee? … only I'm a bit worried about you.' Though responses may vary we need to keep in mind that a *minimum of one in six colleagues need asking!* And the five out of six that don't need it will most probably appreciate it anyway! (That's just a building block of 'culture creation').

The Mental Health First Aider Controversy

In the late 1970s, at 16 years old, I worked briefly at Llanwern steelworks in Wales. I actually rode there on a moped, but I want you to imagine I was dropped off by my mother, who was told: 'Mrs Marsh, it's true we have molten metal splashing about, a range of noxious gases, forklift trucks charging around like it's a grand-prix and a horrible, bullying, toxic management culture – but be assured young Tim will be OK with us as we have no fewer than three highly trained first aiders out there to look out for him …'

She wouldn't have been at all reassured, and the point I'm trying to make, of course, is that a more pro-active and holistic approach would be both more reassuring and effective. So, while mental health first aiders are a welcome and at times highly effective tool in the toolbox – they should **never ever** be seen as a magic bullet and as a substitute for a genuinely holistic approach. (A view articulated by Dame Carol Black at a conference recently).

The best organisations will see wellbeing initiatives and mental health first aid as just two – relatively unimportant – elements of what should mostly be a pro-active, holistic, and integrated approach to human error risk management.

Developing a Culture of Care

As I've said several times already any good safety professional knows that 'culture is king,' and that excellence in culture is objective, analytical, empowering, and dialogue-rich with leadership and emotional intelligence to the fore. Applying a range of practical methodologies based on these principles has enabled the UK, for example, to post world-class safety figures (while remaining very average – at best – in terms of mental health and wellbeing).

We know, for example, that the question 'Why did you switch down?' often means 'You had better have a good reason for that!' But the question 'Why did you think it was safe to switch back on?' assumes you do, with all your operational experience, and so is asking: 'What happened and are we assured it's under control?' It's a matter of *mindset*.

Likewise, in wellbeing, the mental health charity Mind emphasises that we shouldn't automatically say, 'You have to be able to handle stress to work here' as this suggests it's all about personal resilience. Better, it suggests, to have a mindset that says, 'Despite our best efforts, things can get quite stressful.' This points the organisation's mindset at a more holistic and pro-active approach because, ideally, mental health first aiders won't be needed at all.

As Duncan Spencer, IOSH head of advice and practice, says:

> *This culture of care must include systematically identifying the daily stressors caused by operational demand, and putting in measures to relieve that pressure.*

To paraphrase the classic safety differently question, a company that frequently asks its workers 'We want you to thrive and flourish here. What do you need?' won't be going far wrong.

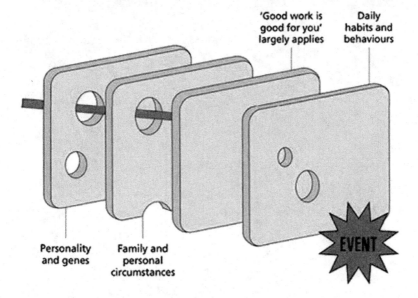

FIGURE 5.2   Adapted cheese model figure.

A Just and Objective Culture

As briefly mentioned above, an objective, analytical, and just approach to error delivers more accurate analysis and a more effective response. This is described in perhaps the best safety book of the last few decades, Matthew Syed's *'Black Box Thinking: The Surprising Truth About Success.'*

An analysis-based approach also motivates and empowers those involved in the process and is central to the concepts of Carol Dweck's work on mindsets and Amy Edmondson's concept of psychological safety. (Note that these multi-million-pound selling books weren't written with mental health in mind, but profit, sustainability, and organisational resilience). They all, however, point at the 'Poke Yoke' mentality of 'always blame the process, not the person.'

In short, where management are in the habit of asking 'what's going wrong?' 'why?' and 'what do I need to do about it?' is psychologically healthy.

6 Soft (or NT) Skills

If I had unlimited resources – I'd not give everyone therapy I'd give all parents and bosses emotional literacy training …

Dr Susie Orbach (One of the world's leading psychotherapists)

Perhaps the easiest and most generic way that an organisation can enhance the well-being of its workforce and the quality of its dialogue and 'care' is by upskilling front-line management in their core 'soft' or non-technical (NT) skills. If culture is king (which it is) and strong cultures are built primarily on learning and excellent adult-to-adult dialogue (which they are) then, by definition, helping management enhance this skill set is key.

For example, (as discussed under neuro-diversity above) the 'Peter principle' suggests that employees are promoted until they become incompetent through lack of support and training. Hence, the excellent but introverted and socially clumsy engineer can easily become an ineffective and highly stressed manager.

A FRAMING CONCEPT – 'Nudge THEORY' OR 'Behavioural ECONOMICS'

The most famous nudge is the painted fly on the toilet in Amsterdam Airport, which men can't help but point at, with a reduction in splashing of up to 80%, leading to an associated improvement in cleaning costs and the environmental impact of cleaning materials.

Crucially, this idea came from the man tasked with keeping the toilets clean.

This, I'd like to argue, is not always mentioned but utterly vital.

Nudge theory, (see the book of that name by Thaller and Sunstein) can be defined as making a change to the environment that is clever, cheap, based on an understanding of psychology and/or physiology, and validated. Just about every interaction where soft skills are employed well is, you might argue, a 'nudge.'

It's a huge topic but a simple example (used earlier but worth repeating) of applying nudge thinking to wellbeing. We know that everything before a 'but' in a sentence will be discounted and everything after the but is the meat of the communication. You're a nice chap and I've enjoyed our two dates but, or, from safety, and infamously 'safely but by Friday' leading to the job being done by Friday as safely as is viable, then claiming 'but I clearly told them safely …' when the investigation starts.

Individuals like the Dali Llama have trained themselves to be in the habit of not saying 'this person is normally OK but today they're really annoying me' as that points us at conflict and judgement. Better to say 'this person is really annoying me today but normally they are perfectly nice' as that points us at analysis, understanding, and care.

DOI: 10.1201/9781003449997-8

THE LIMITATION OF SOFT SKILLS

There's a famous old safety exercise which applies to wellbeing and demonstrates the vital importance of facilitating behaviour – not just with soft skills and 'nudges' but with basic *design*.

In the exercise a volunteer is blindfolded and tasked to throw a handful of marker pens or spoons into a bucket from 10 feet away. In the first run, they get no feedback whatsoever. In the second, they receive only negative feedback, and in the third, all the feedback is positive. They typically do badly in the first run, worse in the second – even refuse to continue if jeered with enough gusto! But they always perform better in the third.

It's said that praise is around 20 times more effective than criticism in changing behaviour and 'catching someone doing something right' is, with reference to ABC theory, a 'soon, certain and positive' payoff. On the third run, a thrower will instantly improve and will typically actually hit the bucket in 6–10 throws. Big cheer! This is the end of the old exercise.

However, this is still, of course, entirely person-focused, and we've added a fourth iteration over the years that demonstrates this. Told that the rules remain that the person throws the pens one at a time, blindfolded and from 10 feet, but that everything else is open to a design or teamwork solution, we soon have funnels and catchers and coaches suggesting a lobbed throw to help the catcher. And we get 100% success. So, while praise may be 20 times better than criticism, it's no substitute for *designed facilitation*. Encouraging people to ride bikes is great but ideally, you'll provide somewhere safe and dry to park them up. (And the CEO will be seen riding in – with any luck fresh from a meeting where they persuaded the local council to upgrade the cycle lanes).

There are lots of ways we can create a culture that encourages and facilitates wellbeing and excellent dialogue naturally with high-impact low-cost behaviours. The rest of the chapter covers the key ones.

LEADING BY EXAMPLE

Back to the CEO on a bike…

It's not really a soft skill of course but understanding its importance is utterly vital. In the world of safety culture, we've long known that any example of a leader failing to follow their own advice is disastrous. With peer pressure, there's a tipping point of around 90% where the new starts and subcontractors will fall under a lot of (often unspoken) pressure to stay with the majority norm. With leadership, however, 90% can be a calamity.

This is a huge problem where management is visibly working themselves into an early grave, unaware of the research showing that working more than 40 hours is pointless (as we pace ourselves and/or need to re-do more things and output remains much the same). If the general culture is brutal, then the best individually-based resilience work in the world is just damage limitation. (The famous psychologist R.D. Laing said that the best way to drive someone insane is to tell them that you love them, then behave as if you don't. So, you can say one thing and model another, but they won't listen to the words *and* it's very stressful!).

A positive example: The CEO of Anglia Water, personally lost 1 ½ stone in weight to show his commitment to their health initiative. He commented that it really added credibility to the questions and challenges he raised in the canteen about the availability of healthy options, and why some things hadn't been changed enough. In America, recently the United Rentals CEO pledged to lose 25 pounds or donate $25,000 to charity.

The CEO also has an important role to play in embedding approaches within the line. Managers may well complain that they have enough conflicting demands to juggle without 'wellbeing' being added 'on top.' It's vital that: (a) It isn't; and (b) that the C-suite makes it clear that this is just a key element of a generic push for excellence and sustainability – and a cost-effective one at that. Then: (c) Ensure it is resourced and coordinated properly.

In short, with wellbeing, as with everything else. Talk is cheap and genuine leadership is paramount.

COACHING

If you understand the feedback fish, you understand the basics of coaching. Imagine your four-year-old has brought you a picture of a fish, but it's nothing better than an outline. Instinctively, you wouldn't exclaim, 'That's rubbish, it looks more like a biscuit, you're wasting my time here.' Instead, you'd say, 'Oh that's fantastic … what a great fish. Thanks.' Giving as much praise as is suitable for the situation is rule one of coaching. (You might want to adjust a 'way to go Joe' USA style positivity on a North Sea UK Oil rig for example).

Then, the second technique is the use of questioning. With your child you'll ask, 'I wonder … how could we make this picture even better? Let me think … how do fish see?' and the four-year-old will shout, 'An eye! We need an eye!' Then they'll add one in. In response to the question, 'Fish swim, don't they?' we'll get fins, and finally a fully formed fish.

FIGURE 6.1

It's the same principle with any coaching conversation. Even though they know you know the answer, so long as they say it first, they will have proved they knew it and will own it. Studies with electrodes monitoring brain function show it lighting up like a Christmas tree when we say it first – *whether prompted or not.* We just don't light up when we say yes in agreement to a suggestion or recommendation from someone else. We may agree, but we don't 'own it' and seldom take it away with us with any degree of enthusiasm.

A trap we can all fall into too easily is to augment their idea or answer with something of our own. Even though, often, it'll improve the idea by say 5% or more, it'll almost certainly reduce their ownership and motivation by a full 50%. So, if they don't ask for input, don't give it. Always keep to questions. Unasked-for advice is so often what's called 'positive but destructive.' It's better than actively negative 'that'll never work' or passively negative 'it could work, I guess, if this doesn't happen,' but none are as good as getting your report to say it first. (You know you've got it wrong when you find yourself commenting later, 'I was only trying to be helpful!')

'Motivational INTERVIEWING' AND THE 1 IN 10 TECHNIQUE

This is coaching in its purest sense, as no advice is given at all, not even any (helpful) leading questions. It all comes from the person, and it's all the more impactful for that. The rules are to ask:

- What would you like to improve on?
- Why would you like to?
- What's stopping you from doing it?

As above, the golden rule is to show patience and not give advice at all, so as to engender a discovered learning/internal promise moment (as above, the best kind, as by far the most impactful). Done well, they'll say something like, 'Yes, I see clearly now what I must do, I can't believe I've let X put me off – out of my way, I'm a woman on a mission.'

If that seems a little optimistic and vague you might prefer this technique.

'1 IN 10' TECHNIQUE

A builder of submarines won a safety process of the year award by having a team of six safety reps tour the site and ask this question of a worker or two once a week. It's taken from educational psychology, where the counsellor tries to build on the answer from a terrible student, 'I'd score myself 1 on a 1–10 scale,' by following up with the question, 'How can we get that up to a 2?'

When someone has said, for example, that they are an 8/10 at driving safely, the key is to ask why they aren't a 0. (Rather than ask why they aren't a 10). Then, when being told about all the good things they do, we can nod and smile, provide positive feedback and generally build rapport. We should have, by this point, minimised defensiveness and be primed to have a constructive discussion. So, we can say perhaps, 'My job as a safety coach, seeking a step change, is to get you from 8 out of 10 to 9 out of 10 and halve those unsafe behaviours. Any ideas?'

In general terms, the same process can be applied to any element of wellbeing or to task re-engineering itself. Effectively applied, it can positively transform your business team by team and task by task.

A famous Stanford University study by Professor Carol Dweck shows that if you praise the effort rather than the innate talent of the person, the results are extraordinary. People praised for being clever will enjoy that, of course, but may show a tendency to avoid challenging tasks in the future because trying and failing might undermine their self-image and esteem. On the other hand, people praised for their *effort, determination,* and *resolve* are far more likely to embrace these traits as, succeed or fail, their positive self-image isn't at threat.

In short, coaching the right behaviours and or mindset can be life-changing.

A sporting example: A football coach might seek to use a bit of 'positive labelling' … perhaps saying, 'Don't worry, Diego, you'll start scoring again soon – you're a natural finisher,' but it's even better to say, 'I really admire the way you keep busting a gut making those far post runs match after match, even though you've had no luck at all for weeks. It's obvious you know from your extensive experience that any match now, you'll get a handful of chances and probably a hat-trick.' He might add, 'I know you're really mindful that the young players coming through look up to you and copy you and that you always try to model the right attitude and behaviour.' (Now he may well walk off thinking, 'I wish we could sell this increasingly useless and lazy old goat and buy someone half decent,' but he'll have done his job as a *coach* at least!)

In summary, the best coaching is where a person we trust (ideally a mentor who models all the things we aspire to) leads the person to set themselves a challenging but achievable SMART goal of their own volition. (SMART being specific, measurable, agreed, realistic, and time-set). Then, they monitor, support, and coach as required.

DELEGATION, COACHING AND FEEDBACK

It's important not to confuse delegation with abdication. Empowerment is largely based on choice and autonomy, but that's not just allowing people to 'get on with it.' Guidance, coaching, or even one-to-one mentoring will also be required as part of a transformational leadership approach. Giving good feedback is central to doing this well.

Imagine trying to learn to play golf, but blindfolded, with no feedback as to where your ball had gone. You'd never improve. How about that exercise bike, treadmill, or rowing machine in the shed you used to use but stopped once the counter packed up? Just like that exercise regime, without monitoring and feedback, we tend to grind to a halt.

BASIC FEEDBACK

When someone is observed doing something constructive, then any praise that is soon and certain will, when delivered from a credible source, be rewarding and will help reinforce the behaviour. The more of this, the better, and books such as Ken Blanchard's *The One Minute Manager* illustrate. (Perhaps the key advice in the book is to try to 'catch a person doing something right').

Whether it's volunteer behaviour, like putting a name down to be a mental health first aider, or where a healthy behaviour has been recently enabled (going to the on-site gym; running an FI toolbox talk, praising a colleague), we'll still want to reinforce it to ensure it embeds and becomes 'what we do around here.'

GIVING NEGATIVE FEEDBACK

By far the biggest issue with negative feedback is either that it's the *only* sort used, or in many organisations, a pronounced *lack* of it. (i.e., 'blind eye' syndrome). This is an issue, of course, because what we fail to challenge, we effectively condone. Challenging can be uncomfortable, so many avoid it if they can. 'It's OK to be challenged and it's OK to challenge' is an expression often heard in an organisation desiring a strong culture and must be a key aim of any organisation seeking excellence. Training people in the skills described here is the easy part. Creating an environment in which they are used readily is the hard bit. (See the section on 'Vroom Marsh Model of embedded training in an early chapter').

Specifically, when giving negative feedback, it's vital to follow some key rules because if you break them, the impact will be the opposite of that intended. They won't be thinking of the behaviour at hand and any associated health or safety risk, they'll be thinking of you, and not in a nice way. If they can, they'll let you know on the spot, usually through their voice tone and body language, but sometimes more directly. If not, they'll most likely just reduce the amount of discretionary effort, which is exactly what we don't want when we're seeking to empower. Often, it'll be both. But more than that, they'll find the experience annoying and *stressful*.

The golden rules of giving negative feedback are:

• Never personalise.
• Never generalise.
• Never berate someone in front of others for extra impact.

A loud 'You're always doing this, you idiot' is a funny line to put in a 'Don't do it this way' training video, but *only* there. We rarely do anything 'always,' and if you suggest that someone does firstly, they'll be thinking that you're wrong. (Even if they do it sometimes and it's important). If you do it in front of peers, they may laugh, especially if it's frequently true. But the person involved will spend the rest of the day fantasising about paying to have you shot. Worse than that, if they're popular and/or it isn't true, the colleagues won't laugh. They'll be mortified on the target's behalf. (If you look up 'workplace bullying' in any HR policy, you'll find that being criticised in front of colleagues will be high on the list).

Here's an example that has stuck with me over the years. The chap involved was an excellent leader: Fair, resolute, consistent, thoughtful, clear in his communications – everything a leader should be. A former rugby league player, he was a key figure in a project that saw accident rates cut to a tenth of previous levels nationwide in 18 months.

One of the things he knew how to do was to give feedback for maximum impact and minimum unintended consequence. The incident the author remembers was at a residential training event where one of the employer's trainees was rude to a waitress. This chap saw this, but said nothing, waited a minute or two, then casually asked the young chap if he could have a quick word about something that had just occurred to him. Unconcerned, the young man followed him out of the room, but when he returned five minutes later, he looked shaken and white. Before he returned to the course, he gave a sincere apology to the waitress.

Normally, tactfully taking someone to the side and sticking objectively to the facts at hand is all that's required. That's exactly what this man did. He didn't personalise, generalise, or raise his voice. But this clearly didn't stop him from articulating his critical observation with impact.

In doing so, he'll have done that rude young man a service. And that's the point. Negative feedback is sometimes required, with even gurus of positive psychology such as Martin Seligman suggesting a 3:1 positive to negative feedback ratio. Whatever the ratio, doing it well is essential. Many individuals avoid the difficulties here by rarely doing it at all, and then, like a backfiring motorbike, charging in unadvisedly all guns' blazing.

BIFF, BIFF (A SECOND TIME), THEN BIFFO AND MAYBE EVEN BIFFOFF

These simple acronyms help anyone remember how to give negative feedback well, especially if we're angry and struggling to operate from that rational prefrontal part of the brain. It stands for:

- Behaviour – this is what you did causing this …
- Impact – and this is why it's suboptimal … so
- In Future, we like this behaviour instead …
- For this (different, more positive) output.

It's BIFFOF because, after repeating it once, if appropriate, but certainly no more than once, we really need to upgrade to:

Or the *f*ollowing will happen …

The second and final F – for *f*ollow through because if this follow up is fudged all credibility will be lost.

*Note: In recent years, some trainers have started to use BIFFO where the final O is for 'open questions' such as 'what do you think about what I've said?' or 'what do you need from me to help you with this?' In short, after BIFF, BIFFO turns to *coaching* whereas BIFFOFF turns to *consequence management*.

ACTIVE LISTENING

You'd think that the skill of listening wouldn't take long to master, and it's true that the component parts of an active listening technique are not difficult to grasp. It's using them in action that's tiring and difficult. And most of us need to practise them! The component parts are as follows.

PAY FULL ATTENTION

This is not just turning off your phone, not looking over their shoulder at something more interesting, or simply cutting them off rudely. Nor is it simply waiting for them to finish speaking so that you can say your thing. It also requires actively concentrating on their words and body language – remembering that typically 85% of communication is in the tone and body language.

This, of course, encourages the speaker, and more than that academic studies show you retain far more even if you're only pretending to listen with positive nods and murmurs and other body language. (The study had some participants pretend to be interested to encourage the speaker … then given an unexpected chance to win money for remembering facts). It's a bit like forcing yourself to smile by holding a pencil in your mouth. Even a forced smile elevates the mood!

PARAPHRASE BACK WHAT THEY HAVE SAID

This is confirming listening by repeating it back in slightly different language, showing you have *processed* it rather than merely parroting it. It's also important to paraphrase back the meaning you got from it if there's anything in the 85% mentioned above that jars or needs clarifying.

This is key. If they sound as if they don't genuinely mean it, then this is the time to clarify. This doesn't have to be confrontational. Something like, 'John, I'm getting the impression that you have concerns about X & Y … could you clarify that?' suffices.

CUT THE BOXING RING DOWN WITH A 'YES/NO' IN THE FACE OF CLEAR EVASION

If you feel you're being fed a mixed message *deliberately*, perhaps to pass the problem onto you, then 'yes/no' is your friend. As above, saying, 'safely, but by Friday' suggests 'by Friday,' so a way of clarifying that would be to challenge by asserting, 'I *can* do it safely, and I *can* do it by Friday, but I may well not be able to do both. What I'm hearing is that, given the choice, you want this out by Friday, come what may?' It's likely that our supervisor will bluster, 'Er … I didn't say that.' Then, to drive home the point, you will need to ask, 'So, to clarify, if I need to delay until early next week to ensure no corners are cut, is that OK?'

You may well get an, 'I didn't say that either!' but they are in a situation now where they have to either say one or the other, or clearly refuse to say either. In any inquiry following something going wrong, the initial 'get out' that they were trying to set up, consciously or otherwise ('I explicitly said do it safely … I'm not sure what the problem is') is now removed. It will be clear to all now that they didn't say anything explicit but tried to fudge it and pass the problem on to you.

Of course, allowing people to pass their problems to you is stressful and bad for your wellbeing. This overlaps with assertion.

ASSERTION

Assertion is not to be confused with asserting yourself as 'insisting' on something for you. It's merely insisting that your rights be respected without impacting on those of others. Being able to give negative feedback well and to challenge mixed messages – as above – are key skills of assertion. For illustration, an assertive 'argument' might go something like this:

With respect, boss, I think X decision is wrong. What about Y factor?

That's a good point, and actually considering factor Y did indeed give me pause for thought. However, on balance, I still think we should do X, and I get the casting vote.

OK, I'll try not to say 'I told you so' if you're wrong. But that's not a promise.

Understood!

The key aspects are mutual respect, and that the boss still gets to be the boss. This is an 'I'm OK, You're OK' interaction from the classic Thomas Harris book of that name.

Of course, mutual respect might be thin on the ground. In this case, techniques such as 'broken record' and 'paraphrasing the message' can be used to show you've listened and understood, without agreeing with it. Broken record is where you stick to the issue and refuse to move on until the person responds – or at least clearly refuses to! (See 'yes:no' above). As many a political interviewer has quipped, 'that's a very clever answer but, for the record, it's not exactly a "no I didn't" is it?'

A paraphrase example: Imagine a really difficult client who is proving impossible to work with even though there's little profit in the contract, and who will prove even worse if you tell him you're not going accept the next job he wants you to do. (They'll find an excuse to withhold payment for this job for a start.)

Among other gripes, they warn you that putting the price up may cost you any future contracts, so you say, 'I hear you, price is key, and putting the price up may cost us the next batch of work. Thank you, I've noted that.' (They'll be thinking 'good' but you'll be thinking 'though it's utterly irrelevant' – though keeping that to yourself until the time is right.)

CASE STUDY TO INTRODUCE TRANSACTIONAL ANALYSIS

Here's an example of a company that took that approach to non-technical skills and made it the centre of their entire culture enhancement strategy. It was called 'Middle Bubble Training' after transactional analysis (TA) theory, which says we can be in one of three states when interacting with others

- Parent;
- Adult; or
- Child.

The child state is only appropriate in a brainstorming session, but the parent state should also be avoided as much as possible. The style of an authoritarian parent shouldn't be confused with a directive leadership style, which is appropriate when employees are inexperienced or at risk, but often constitutes overuse of a top-down

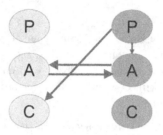

FIGURE 6.2   Middle Bubble Training

'because I say so' mode. At its worst, an authoritarian style will almost inevitably engender a 'balancing' negative response, which we've discussed above, and which is a disaster when trying to deliver an empowered, respectful, and caring culture. However, even the more benign 'nurturing parent' style has two negative side effects.

First, if you assume you know best, it will inevitably hinder your listening and learning. Second, any sort of paternal or maternal mindset won't be likely to do the worker's empowerment any good.

The 'adult' state in this case is essentially about assertion, active listening, and giving good feedback, especially when that is negative. It's very easy to think of things in threes: So, top bubble, middle bubble, or bottom bubble? And the key principle is simple: The more often all employees are in their middle bubble, the better.

Unless you are a trained killer or some sort of sociopath, it's really difficult to engage that prefrontal cortex, *think*, be in the middle bubble and *react* simultaneously. For most of us, they are mutually exclusive and you can't think and react simultaneously. How many of us have blurted out something without thinking and then as the front brain catches up instantly regretted it!

Here's a true example of how user-friendly the model is. 25 years ago, your author still occasionally frequented places where people congregate to drink alcohol and listen to music. Late one night, a huge man tapped the author on the shoulder and said, 'You just spilled my pint.'

'I don't think so,' your author replied, looking up. 'Yes, you did,' he said, leaning over intimidatingly. Trying to stay calm, I said, 'Look, I really haven't bumped into you or anyone else. You've got the wrong bloke' while thinking 'where are the bouncers!?' Looking the author in the eye from too close for comfort, he insisted, 'No, I've got the *right* person,' but then with a giggle, 'but luckily for you I'm in my middle bubble and all's good! Ha ... had you going there a bit Doc!' Thankfully, I recognised a trainee from a behaviour-based safety course run a few years before; a 6-foot 4-inch forklift driver with, I recalled, profound literacy issues. I mention this man's literacy issues only to stress how easy and appropriate it is to train everyone in interpersonal techniques that can sound complex in a textbook but are actually quite useful and user-friendly.

I'd like to argue such models shouldn't only be reserved only for management and a very well attended immersive safety day uses a version of this model – referring to it as the 'three circles.' In this version, the bottom bubble is where the person doesn't want to communicate at all, the top where management (or colleagues) communicate 'broadcast' like a radio beacon only with no listening. In both cases, the phrase 'behaviour breeds behaviour' is applicable where it's stressed that if we stay calm, listening, and keen on dialogue then often the person wishing to broadcast only, or avoid all communication will, in time, mirror us.

Such generic skills facilitate the effective flow of analysis and communication, both of which are utterly central to creating an empowering, blame minimising, and stress-free culture. They clearly fit under the 'nudge' banner, as they cost little but can have a big impact on behaviour.

In short, if we really want to create a culture where 'it's OK to be challenged and it's OK to challenge' that is psychologically safe, we need to equip individuals with the skills to do that. Insisting they do it without training is, again, well, very stressful.

Talking about psychological safety, in her book 'The Right Kind of Wrong' Amy Edmondson stresses the vital importance of learning (which this book is all about) and, among other key issues, the importance of owning our own mistakes. Perhaps the best technique for this is "regret, reason, remedy'. Here we apologise for what we got wrong, explain and say what we'll do to make up for it. (Of course all with sincerity, insight, objectivity and integrity! But that's another book).

Of course we can get defensive, pretend it didn't happen, scapegoat someone else, seek to 'reframe' it like a slippery politician - or just pretend to faint like Basil Fawlty. But all of these are sub optimal for the company culture and for your own reputation compared to RRR.

DIRECTLY ENGENDERING OWNERSHIP – TWO CASE STUDY

I've tried to describe some day-to-day habits and skills that can empower employees. Briefly, I'd like to describe two more direct case studies.

First, an old friend and colleague, Jim M, oversaw the doubling of turnover of his construction SME by applying a dialogue-based approach looking first at the stacking of materials in the engineering workshop, and the loading and offloading of vans.

Each suggestion made the job a little quicker and easier to do, but the real benefit he suggested was a massive increase in motivation and engagement. He proudly showed us pictures of pieces of kit that looked like giant toast racks and giant trays, all made by his engineers in their workshop. He conceded, 'They may look a bit rough and ready, but they're perfect for our guys' needs and they're really proud of them.'

It really is as simple as that.

Another excellent example of genuine ownership of a behavioural-based safety process would be a shipbuilding company that offered, at process launch, a weekend for two in Amsterdam for a winning logo design. Many entries were received, and the prize was won, but the workers weren't impressed. 'Every day this ship is delayed costs them millions. What's a weekend for two in Amsterdam?' complained one employee, while others went further, alleging that the incentive was 'just a bribe, really.'

However, it was noticed that 11 entries were from worker's children, and, with a twelfth quickly 'commissioned' from the CEO's son, we had ourselves a calendar. This *did* land well and needed reprinting twice. With 2,000 workers on top of each other trying to get the platform finished, the job and the deadline were challenging, and the ship was only finally commissioned while being towed to its initial mooring, but there was not one single lost time injury (LTI).

It's about the psychology of the leader-member exchange, and that's about perception. Management can't tell workers they have ownership or to be motivated; workers tell management they are – or are not. This is, I argue, as much to do with one-to-one exchanges and soft skills as it is about the formal delegation of roles and responsibilities.

REDUCING AMBIGUITY AND STRESS

It's worth briefly covering a few key techniques from the world of influencing skills. In this context, they're not about influencing people to do what we want them to do, but to set a tone of competence and integrity with the aim of communicating clearly and reducing ambiguity and stress. They are:

- First, the one you heard when you were about knee-high (to quote the song). The basic polish your shoes, stand up straight, look the world right in the eye and speak clearly.
- Speak decisively. Don't say 'if I can' or 'I'll try' unless you absolutely have to.
- Never, ever say, 'I've been told to come and tell you …' (Either do it right or say no to the person who tasked you with passing the message on!).
- Treat people like adults by using data and illustration to make your case. We've all resented being told 'because I say so' since we were 8. (The self-fulfilling prophecy of Theory X and Y shows that people who are treated like adults tend to act like adults, and vice versa).
- During dialogue, keep a neutral and open body position. Palms open, hand gestures natural and gentle. No folded arms, no pointing or chopping.
- Remember that humbly saying 'you're the expert' sets you up to be trusted, as you are at your most persuasive just after you've admitted a weakness (and trust development is our number one aim).
- Use the 'I' word, and get the person you're talking with to use it. We are something like three times less likely to break a promise if we use the 'I' word in a sentence. (For example, if, when visiting a beach, someone asks you to please keep an eye on their bags while they take a swim and you look them in the eye and say, 'I will,' you do. On the other hand, if you ask someone to look after your bag while you take a dip, and they mumble 'sure' while looking at the ground … you don't go swimming!)
- Use people's names. It's the sweetest sound in the world to even the humblest person.
- Finally, take a tip from Bill Clinton who was a master of this. Almost everything, including perceptions of integrity, goes better with a little bit of touching. (No giggling at the back – I mean appropriately firm handshakes, appropriate pats on the back or shoulder taps).

People really skilled at these that have no integrity, such as some politicians and other sociopaths, can still be really influencing, but they very often leave you with a sense of discomfort that defeats the object. You say yes, but you feel suspicious and uneasy and almost certainly *stressed*. (It has to be conceded these people can achieve a lot in life – with the exception of formal inquiry findings, and reputations and biographies their families are proud of).

It's far better to use these influencing techniques to reinforce the fact that really *do* value their health and safety

CONCLUSION

VICIOUS AND VIRTUOUS CIRCLES

Finding yourself angry, ise ideal. Bad work is bad for you and it's easy to imagine a row, some self-medication, and a bad night's sleep being more likely. The flipside is that good work is good for you, and people who enjoy and get meaning from their work can often be psychologically healthier than those who don't need to work at all. So, we return home smiling, and virtuous circles of all sorts are more likely to follow.

It's long been accepted that a good walk and talk allows for the exchange of ideas and shows visible felt leadership and commitment. So long as the management involved are genuine in their intent, listen and don't just file any good ideas in a drawer, it's a key element of any cultural enhancement approach.

However, there's also something rather more self-fulfilling and subconscious going on that underpins both vicious and virtuous circles. I'm returning, of course, to the whole issue of fast or back brain function and it all boils down to two cocktails of chemicals.

The first cocktail is comprised of *adrenalin* and *cortisol* and these chemicals occur when we are, for example, *stressed*, *uncertain*, *anxious,* or feel we have been treated *unfairly*. The problem with these is that they automatically increase …

- Intolerance
- Irritability
- Criticality (blame)

And reduce or impair…

- Memory
- Creativity
- Decision making

And of course, lead to less discretionary effort, worse mental health, and more incidents and accidents.

Not really ideal under any circumstances. On the other hand, the more joyful cocktail comprises of *dopamine*, *oxytocin,* and *endorphins* and these are generated by such as *humour*, *engagement* and *empathy,* and perceptions of *fairness*. These promote:

- Calmness
- Focus and attention
- Creativity
- Generosity
- Bonding
- Trust

And, of course, more discretionary effort, better mental health, and fewer incidents.

I think we can all agree that that's a bit more like it when we're seeking the multi-faceted benefits of culture and resilience creation – with many organisations very explicitly striving to build a 'culture of care' as well as one of safety and health. Done well, we can add a blank slice in that Cheese Model, break the chain, and help prevent something awful.

Soldiers, for example, have relatively normal suicide figures, according to the UK ONS, but that's not true of ex-soldiers who find they miss the camaraderie of their profession. To refer to the adapted cheese model when soldiers hang up their uniform, it can remove an impenetrable cheese slice. So, especially if they take some untreated PTSD with them, when they leave their jobs, the prognosis can be bleak.

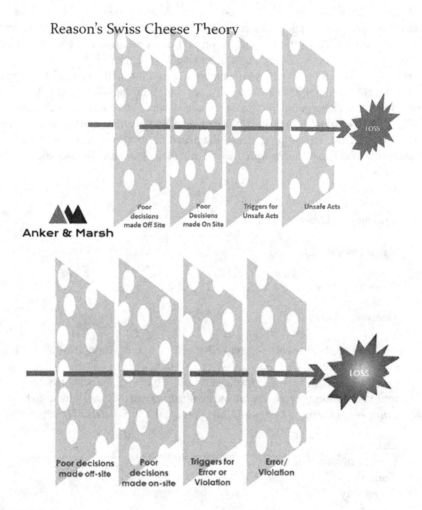

FIGURE 6.3

However, although we can't all enjoy the camaraderie of military units, we can share their professional satisfaction. We can work in organisations where task-based issues such as the amount of autonomy and control and skill development are tweaked to best suit the individual and daily dialogue is objective, adult to adult, and constructive.

Wellbeing, Motivation and TVs 'Undercover Boss'

If you've ever watched a version of the international TV show Undercover Boss, you'll have noticed that the boss always learns a lot and nearly always develops more respect and affection for their workers. It's a nice user-friendly validation of the techniques referred to here. However, it's also a variation on the old truism that it's impossible not to have at least some affection for a person whose story you know, and that links back to the holistic nature of wellbeing and motivation.

Most programmes finish with a worker being promoted, and they always beam with smiles and announce they can't wait to tell their family. Often in these shows, the boss will go on to say, 'I'm just so thrilled to have someone like you working for me,' and very often the employee will well up. More money is always really welcome after all, but that's a practical issue.

Just like emotive music, being valued as a person gets you straight in the primeval reptilian brain. In practical terms, it means you're safe from being pushed out of the cave anytime soon. Considering that we upgraded our brains millions of years ago to include the prefrontal cortex, it's amazing how many thoughts, behaviours, and emotions are still controlled by that original old laptop! Am I safe? Am I useful? Can I trust you to have my best interests at heart?

In short, a good wellbeing process will be holistic and address the whole person – especially their subconscious. Having the right time and tools to do the task in a safe manner is a start, but the benefits of 'good work is good for you' require systematic consideration of the whole person. The American activist and writer Maya Angelou has a lovely quote.

People will forgive and forget what you did and what you said but they never forget how you made them feel.

And these feelings plug directly into key KPIs like turnover, absenteeism, presenteeism, discretionary effort, mental health, and accidents. In short excellence in health, safety, and wellbeing is the very definition of win-win.

As we emerge from COVID-19, with every fatigue and mental health metric heading steeply in the wrong direction, it's in the best interest of all organisations to *pro-actively* address these issues. Work may be only one element in the equation – and sometimes only a small element – but on many occasions, we can help break the chain of vicious circles and set up virtuous ones, and that can make all the difference in the world.

Section II

7 Planning and Design

I've argued throughout this book that almost anything and everything to do with an organisation is applied BBS.

As briefly mentioned in an earlier chapter, one of the most influential occupational psychology papers of all time is Schneider's ASA theory, which says that certain people will be *attracted* to an organisation, while others perceived a good fit will be *selected* by it, and finally, there will also be selective *attrition* as the accuracy of these perceptions works through. We have constantly stressed that everything contributes to safety culture, and 'culture is king' so to illustrate this principle, I'd like to consider the way individuals are selected, developed, and retained, as 'everything contributes' starts even before the employee joins the company.

Imagine a safety orientated A star * candidate in an interview with a company with a dubious safety and worker welfare reputation. If they are asked solely about productivity and profit by a panel, this will confirm the reputation and they are likely to accept instead a similar financial offer from a more enlightened organisation. If they do join and realise it is as bad as they feared or find they can't make the changes they hoped they could, they will be likely to move on far more quickly than we'd like. So, in addition to basic reward and recognition issues, even the wording of a job advert contributes to the safety culture.

(The reverse is also true. In a company with a strong culture, a person with a dubious attitude will stand out like a sore thumb and will either get uncomfortable and/ or feel 'persecuted' and choose to leave. Or, if they are invited to leave it's far more likely to look a legitimate management response, and perception is key here as a person will have to be pretty bad for colleagues to go against one of the great unwritten 'laws' of peer solidarity and disown them and not 'have their back').

Constantly setting the right tone and throwing the right shadow is arguably as important as challenging unsafe acts on a day-to-day basis, starting with the fact that get that right and there will be fewer unsafe acts to challenge. In short, attracting and retaining switched on and safety aware employees is, among many other similar issues, an element of a holistic BBS approach.

But, that point made., to return to basics.

PLANNING FOR BEHAVIOURAL SAFETY

What we need is to make a practical and viable plan and ensure through monitoring that we see it through to fruition. This might be referred to as 'governorship' and there are 1,001 variations on this theme. Whatever the design, something systemic and holistic is required.

Many BBS processes will include some version of a guiding committee, including safety or other shop floor representatives, experts, and line management.

DOI: 10.1201/9781003449997-10

Often a dedicated project manager or 'champion' will be designated with, ideally, some combination of the two.

What we are describing here is basic change management: Plan, do, review, and re-plan. It's no more complicated than turning a genuine desire for change into some sort of logical plan, resourcing it then assessing systematically and competently how well that plan is working out. 'Our aim is zero accidents, as all accidents are preventable' is fine as a *principle,* but if that's all supervisors get told, it's simply not going to happen. Likewise, a workforce left in tears by a moving personal testimony from somebody who has been injured at work is primed to do something 'better' but simply saying: 'Think on, take care, and don't be like this man' is far too vague.

We might cross-reference 'plan do' with something as simple as a SMART goal here to give us a robust plan. SMART has various versions, but mine is:

- Specific,
- Measurable,
- Agreed,
- Realistic & reviewed and
- Time-set.

Firstly, what specifically are we aiming to achieve? Perhaps a reduction in LTI accident rate from 2.0 to 1.0 within 18 months? This might well be needed for moral reasons because someone has just got hurt and we must address a collective desire that 'something must be done' to prevent that happening again. Or perhaps it's for business reasons as a major client is *demanding* that we describe a realistic plan for substantial improvement. (Top-tier clients, of course, don't like it if you're vague or if they follow up and find nothing has changed).

Ideally, it'll be because the organisation has chosen to pro-actively deliver a win-win without this being on the back of an incident or because they've been told to. However, being cynical, it's not all that often, and even then, usually on the back of the arrival of a new CEO shocked at what they've found. (Though increasingly, post-COVID especially, safety is being seen as a subset of wellbeing, and more and more companies are looking at KPIs like turnover, absenteeism, presenteeism, and discretionary effort as crucial to their very survival. It's a great opportunity for proven safety excellence to be part of the survival solutions).

Cynicism aside, whatever the motivation for halving numbers of incidents, it's something that can realistically be achieved in 1–2 years or sooner if conditions are favorable. Halve it again, then a *third* time over about 5 years, and you have a case study to grace any website or book chapter. If that sounds simple, that's because describing it is. Harder to do, of course, but countless organisations have achieved it, and (with suitable management commitment) yours can too. (There are several international case studies on the Anker and Marsh website).

A practical plan will include a road map, developed by a strategy team made up of operations, HR, HSE, and (ideally) union or shop floor representation that comes up with something as simple as:

- Training all management and supervisors in 'risk literacy' and 'soft skills for safety' over six months.
- Setting up a formal 'walk-and-talk' approach that allows these skills to be applied to safety issues on a weekly basis. *(*Arguably the one thing you do if you do nothing else)*.
- Embedding these new skills through a HR-run(?) 360 follow-up formal appraisal, as well as informal day-to-day discussions and assessment using hand-held check sheets.
- Including an element of coaching to support and encourage people as they try out their new skills. (Or admonish them if they haven't tried them out).
- Training several BBS workforce teams tasked with analysing a handful of problematic behaviours of their choice with the task of coming up with high-impact, low-cost solutions as well as high-impact high-cost solutions.
- Publicising the changes that flow from these teams to maximise the opportunities for praise and to create a 'can do' atmosphere. (Celebrating what the management guru Kotter's terms 'short-term wins').

That's my suggested *essential* list. The following are *desirable* if appropriate and/ or viable:

- Generating volunteers for a peer-to-peer 'walk-and-talk' approach which may or may not involve measurement (as is appropriate).
- Inviting these volunteers to form a committee that will give itself a name (for ownership purposes) and who will drive and control the BBS data monitoring and feedback processes.

This list is a solid top-down *and* bottom-up design that covers all points of such as Kotter's famous change model. There are of course many variations on it. For a start, towards the end, desirables will apply only to a reasonably static and accessible workforce, such as you'd find in a factory. However, since your process is going to be primarily based on *analysis* and *empowering leadership,* then observations (by management and/or peers) are *desirable* but not *essential* and something holistic and systemic will be entirely possible regardless of the challenges.

A few strategic decisions are worth outlining here, since we're talking about overview strategy and it's difficult to distinguish between what's effective and what's symbolic as process and management commitment intertwine. For example, when a volunteer team comes up with some 'high-impact, low-cost' ideas, we'll of course put them into action asap as they will directly improve safety performance. However, maximising the publicity associated with the changes helps create a feel-good factor that will increase the number of volunteers for anything else you're planning to do. It also motivates the original volunteers to go out and find *other* ideas. It's a virtuous circle of building trust and traction, driven by management commitment.

I genuinely believe that whether a culture is improving or drifting backwards is as simple as how many virtuous or vicious circles we're generating.

ESSENTIAL AND DESIRABLE ELEMENTS WHEN DESIGNING A SAFETY FRAMEWORK FOR AN *INTERNATIONAL* ROLL-OUT

If an organisation has a very decentralised approach, then applying a highly-centralised strategy will almost certainly hit resistance, no matter what. However, while local ownership is always a good thing, there are a few key factors that must be the bedrock of any strategy, no matter how decentralised we'd like it to be. Specifically, I'd argue that there are three 'essentials,' requiring a formal plan and associated rationale, *regardless of where in the world you're working, what sort of an organisation you're working with and what the behavioural problems are.* The rationale and elements of each were described above but worth repeating in summary.

- **Learning:** How will the process maximise the objective understanding of why unwanted behaviours are occurring? Answers will differ around the globe, but the more objective the understanding, the more likely (blind luck aside) that the chosen response will work.
- **Transformational Leadership:** How will the local organisation enhance and support the 'soft skills,' such as coaching that maximise the quality of dialogue and the number of 'willing followers' and the associated discretionary effort? Which takes us to …
- **Empowerment and Ownership:** Implemented well, the two issues above will automatically enhance this area. If we coach well, analyse objectively, and lead well, generally this will lead to workforce empowerment. However, we can also empower directly by delegating control of some aspects of the safety process; for example, setting up a committee with a budget that can be applied as it sees fit to any suggestion it generates, even an expensive one. (Though probably only if that's ratified by management as a cost-effective investment in the medium to long-term …).

WHAT MAKES UP A CULTURE?

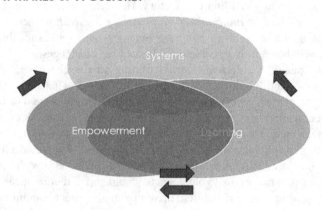

FIGURE 7.1

NB You may be aware of the term 'safety one' and 'safety two' and many expert commentators (including Eric Hollnagel himself) saying that it's 2 as well as 1 not 2 instead of 1. I'd suggest this a user-friendly visualisation as "S1" is essentially the top of the three bubbles and "S2" the two below.

We discussed in the previous chapter how culture is king, and my model of culture seeks to break culture down into just three simple elements. First, systems. The better they are and the easier to understand and use they are – the better they facilitate safe behaviour. If we haven't hit diminishing returns here, then we may well need to work primarily on this area – especially if resources are scarce. (There's a market for 'readiness for BBS' assessments, but this is the only issue for me. If you *have* hit diminishing returns, or close enough to that to have a multi-faceted approach, then most everything else described in this chapter is as applicable as it is essential. And if it's essential, then don't pussyfoot about!)

Facilitating good behaviour should always be the number one aim. However, most companies reading a chapter like this will broadly have hit diminishing returns and have plateaued safety performance. (Over the years, the safety wave has been seen to apply to many, many companies. Typically, they look at their historical safety figures 'waving' and realise two things. First, that many comparable organisations are doing better and that what's left for them is mostly 'behavioural'.

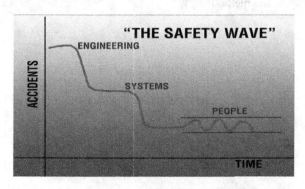

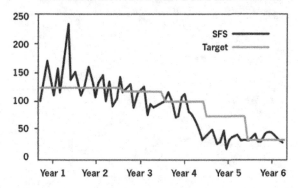

FIGURE 7.2   A diagram of the safety wave as a concept and a diagram with real data from a client to illustrate.

The learning element drives continuous improvement as we address the human factor and the inevitable unintended consequences of change and growth. Finally, everything goes better if the workforce is switched on, engaged, alert, positive, and motivated.

Branding the Process? – A Case Study

Many years ago, I gave a talk – based on the above – at a conference in Barcelona and was approached by the global head of safety of an international oil company and his team, who over the next three years would cut first aid, accidents, and fatalities to a sixth of previous levels. They said they'd had a 'story telling' consultant in, who'd had senior management write a letter to all employees in 5 years, congratulating them on the stunning success of the 'Live!' project they were about to launch. 'Live!' had all sorts of sexy and impressive branding and merchandise ready, and I asked if they wanted me to do a SWOT (strengths, weaknesses, opportunities, threats) review of the proposed methodology behind it before the launch. 'No' they said, they only had the branding and the letter to date and no methodology at all yet. So, could I perhaps go through the talk again … only rather more slowly …

As above, they did it right and reaped the benefits, and the stylish, omnipotent branding helped. But the best branding in the world won't do you much good if it isn't underpinned by sound methodology.

8 The Essential Methodologies

A DETAILED LOOK AT A SUCCESSFUL 'FELT LEADERSHIP' DRIVEN BBS APPROACH

The one methodology that is broadly agreed to be 'the one thing an organisation should do if nothing else' is the 'walk-and-talk.'

This is the (largely) top-down behavioural approach often called 'Felt Leadership,' which appreciates that safety has to be an on-going *process* led by line management leading from the front. This by investing the time to go out onto the workplace and engage the workforce in conversations about safety. (Ideally, they'll have learning focus, will use coaching to enhance discovered learning, and of course, while always lead by example).

There are numerous variations on the exact definition of Felt Leadership, depending on the consultant (or tailored in-house methodology). At its core is a commitment to turning something like DuPont's core principles of safety management into reality through day-to-day leader-worker interactions. These are:

- All injuries can be prevented.
- Everyone is responsible for safety (*and* everyone is valuable).
- Safety is a condition of employment.
- Never stop checking how you're doing.
- Everyone has the right to challenge anyone – and expect action.
- There are no minor injuries.
- Workplace safety is only half the story.
- Never think you can't keep improving.

WHY SAYING 'DON'T DO THAT' CAN OFTEN BE A GOOD THING

We are clearly moving from '*you* make sure you're safe' to '*we* need to make sure it's safe.' It's an important distinction. The Bradley Curve, devised, in a moment of inspiration just before a key meeting, deep inside DuPont by a Vernon Bradley (in conversation, I understand, with Peter McKie), suggests that organisational culture moves from **dependent** ('I act safely when I'm being watched') through **independent** ('I act safely even when I'm not being watched') to **interdependent** ('I am my brother's keeper'). It's a conceptual model that resonates as face-valid around the world, but the linear nature of the model can be challenged. In fact, experience suggests that most employees will start as independent but become dependent and disempowered through factors such as learned helplessness in organisations where an organisation's culture is weak.

DOI: 10.1201/9781003449997-11

"Bradley Curve" ...

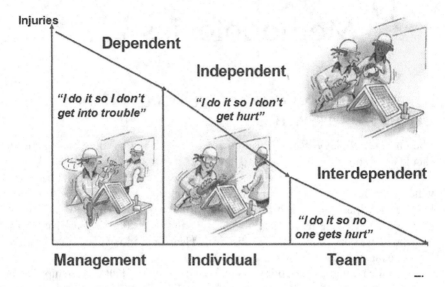

FIGURE 8.1

Regardless, any approach like this gives us a chance to learn directly from the people undertaking the work and the inquiry into the Columbia Shuttle disaster gives a good example. The chairwoman of the mission management team was asked what she did about dissenting opinions. She replied confidently that, when they were brought to her, she took them on board. But when she was asked what she did to find out about dissenting opinions that *weren't* brought to her (and which she, therefore, didn't get to hear about), there was an infamous silence.

Being out and about, we also get the chance to actively coach, praise, and lead by example. But there's something almost as important that's squarely in the unspoken, subconscious grey area that this book seeks to address. It's because it turns strangers into casual acquaintances and the relationship between acquaintances is much stronger than one between strangers. The civil rights protestor Rosa Parks is a good example. She wasn't the first person to be arrested in Montgomery for refusing to give up her seat but she was the first well-known person in town to do so. The dozen or so others were processed and released, but Rosa was on many committees and made dresses for the wealthy elite, so when she was arrested, lots of calls were made and people mobilised.

Indeed, in very many cases, we behave far better in front of acquaintances than friends and old colleagues as 'best foot forward' applies! (Again, driven mostly subconsciously).

So, it follows that or a BBS program to work well, and for the benefit of a safety culture, we want an organisation made up of acquaintances. Walk-and-talks really

FIGURE 8.2   Picture 1 – Rosa Parks arrest. (See below if we can use it)

help with this and mean that people are far more likely to open up to us when we ask them about realities, or even pro-actively seek us out. An ever-present questioning attitude is also far more likely to happen when you're with an acquaintance than with a stranger.

Similarly, the concept that it's okay to challenge and to be challenged works better too. An organisation claiming that here 'challenging and being challenged are valued in this organisation' is one thing but for it to be a day-to-day reality, a lot of ground work needs to be done and a lot of bridges need to be built so that these behaviours become a habit.

WALK AND TALK – THE POTENTIAL WEAKNESSES

So, a walk-and-talk it really must be but now I'd like to address what can go wrong with this walk-and-talk BBS methodology? Or perhaps, where could it go more right?

At its worst, a walk-and-talk methodology can descend close to farce if an overly paternalistic style is adopted. I once reviewed a training course for a client where a video included as one step of the methodology: 'Bob will now use the technique of listening to Billy.' (I'm pretty sure that the video said this was its 'patented technique of listening. TM pending).' I watched it on an oil rig with a group of workers.

Those that weren't openly giggling ranged from the astonished to the positively *affronted*.

A second problem that naturally flows from an overly paternalistic position is a reinforcement of the status quo. In another popular DVD, a manager talking sensibly about workforce empowerment from an office *on the factory floor* with workers busily working away over his shoulder, uses the expression '*down* there on the floor' when referring to the workforce. Later, someone else in the DVD suggests – with a straight face – that 'the workforce may know how your systems work in practice even better than you yourself.' (You *think*!?)

A Step-by-Step Critique

I think it's illustrative to walk through the classic, successful, and comprehensive nine-step SUSA approach step by step, as described in the film *Safety Watch (Outtakes 1996)*. (SUSA stands for *Safe and Unsafe Auditing* and was championed by the consultant John Ormond, formerly of ICI and now retired).

This analysis is informed by the many Strengths, Weaknesses, Opportunities, and Threats (SWOT) analyses of previous behavioural and cultural change programs based on STOP, SUSA, or an in-house version of them that I've undertaken over the past 30 years. You may not agree with me – but that's fine. I may be wrong! But I do hope that if and when I am the debate proves useful and clarifying.

Following this critique, I'd like to suggest a five-step walk-and-talk approach I feel addresses the issues I raise step by step. Finally, I'll return to some contextual problems seen with any sort of walk-and-talk approach.

The Basic Nine-Step SUSA Model:

1. Stop and observe people.
2. Put people at their ease.
3. Explain what you're doing and why.
4. Ask about the job – what are you doing and what are the stages?
5. Praise aspects of safe behaviour.
6. Ask 'What's the worst that could happen and how?'
7. Question 'Why?' of any unsafe behaviour.
8. Ask what corrective action is required.
9. Achieve a commitment to act.

This is clearly a very thorough approach. However, while there's clearly a lot of very good stuff in there, I'll argue that it's rather *too* thorough and a bit top-down. I'll try to justify those observations by addressing the logic and practicality of each point as we walk through. As above, you may not agree with all my points, of course, but I'm hopeful you will find the process useful.

I'll often talk of 'nuance' as I go, so it's worth restating: It's estimated that 90% plus of our thinking is subconscious, and when it comes to culture and, especially, when a worker is talking about something important – they are hugely attuned to nuance. The impact we are trying to make and the impact we are *actually* making may not be the same thing at all. Whatever we do needs to be sincere – and keeping it simple and faced in the right direction really helps with that.

Stop and Observe People

This element stresses that observation must be active and will require concentration, hard work, effort, and some dedicated time. The analogy is with 'active listening' which done properly is actually, rather hard work and the famous basketball scene ice-breaking exercise also applies. (See 'theinvisiblegorilla.com/videos' for a simple example. This is the scene where delegates are challenged to count passes between intermingling basketball players, some dressed in yellow, some in black. The real question is whether the large gorilla that walks across the scene blows a kiss or beats his chest. The majority of delegates, distracted by the players will ask 'What gorilla?')

So, we need to go and look specifically and ideally *only* for safety issues. ('Watch this scene and note what any large animals that wander onto court do' is an easy task!)

Note, however, that this isn't an essential element if the primary focus is dialogue and learning. Indeed, with peripatetic workers where you catch up with them in a canteen/yard, it's not possible.

Put People at Their Ease

Do this by introducing yourself and explaining who you are, what you're doing, and why. It's considered vital that you come across as friendly, concerned, and constructive. You establish a rapport and that you're there because you care.

This is my first area of concern. This principle can't be faulted but this is the first hint of *'paternal* parent' in assertion terminology. Done badly, and it's easy to do clumsily, it can come across as a little ominous (as in there's a 'but' coming up here) or even a little patronising. 'I want you to know I'm here for you' is far easier to deliver by an actor in a Hollywood film than by a badly trained and socially clumsy supervisor on a North Sea oil rig.

It must be said that the delightful and sincere Ormond *himself* always got away with it, of course, but the problem is that someone like John doesn't lead every interaction!

So, I'd suggest the answer is putting people at their ease by being natural with them, rather than making an *effort* to do it.

Explain What you're Doing and Why

It's said that the average person can only handle seven items at any one time. And, for many of us, the spread of plus or minus two means we can only manage five or six. Personally, as I age, I find myself more and more drawn to the use of triptychs – trios.

Regardless, *no* list of this sort should ever be longer than seven and this for me is the first element of repetition that needs to be amalgamated. For me, it's merely the non-controversial element of point 2.

It's also the first opportunity to say (again) that when we're talking about two busy people having a conversation the simpler and more robust the model the better.

Ask About the Job – What Are You Doing and What Are the Stages?

Here it's stressed that the questions should be open-ended rather than closed. We want the individual to talk in depth and openly – not retreat behind closed answers. For example, asking: 'Are there any safety aspects of this task that concern you?' is likely to get a 'no' from someone wary of your motives and/or under time pressure. '*What* are the safety implications of doing this task?' is a better question, followed by 'Which aspects concern you most?' or something along those lines.

This is good stuff and an excellent way of breaking the ice and starting on some *analysis*. It also resonates with the Safety Differently 'what do you need' approach.

Praise Aspects of Safe Behaviour

Again, absolutely! However, this is another area where the question 'but how skillfully?' is raised. For example, in the *Safety Watch* DVD, Ormond stresses the importance of not being phony, stiff, or formal 'like in the old training films' but then gives the example: 'I noticed you were lifting with a straight back … you're not going to be one of those with a bad back' (Even John himself sounding a little stiff and formal!)

Again, someone as passionate about safety as John himself is will get this right/ get away with it, but it is, I'd argue, uncomfortably close to the 'old training films' John himself warns against that so often goes down so badly. Similarly, and to an extent infamously, many upbeat 'way to go Joe' US training films simply didn't travel at all well in the early days of imported to the UK and Europe BBS.

(Here's a challenge to illustrate the nuance I'm stressing here: Read the Australian book 'Don't Tell Mum I Work on the Rigs: She Thinks I'm a Piano Player in a Whore House' by the Australian oil worker Paul Carter. The tales he relates are hair raising but an excellent example of the daily realities many workers face and the cynical black humor they so often use to get through the day. The challenge is: How careful would you be to get the pitch and nuance *exactly* right if you knew Paul and his colleagues were in a safety session you were running?)

To address the challenges here I'd like to suggest three things. One, get everyone to read the Carter book! (I certainly make sure any consultants I employ do!) Two, I'd suggest that whatever process you choose incorporates a praise technique selected consciously to be best-suited to your local culture. Three, and perhaps most importantly, I'd like to suggest that this praising element *follows* any analysis and comes at the end of the session, so that rapport is more likely to have been built. It's just so much easier to say *naturally* something like: 'That was really interesting. You've given me some real food for thought here. Thank you for your time' at the *end* of a discussion than in the middle. It's easy to imagine a conscientious 'walk and talker' thinking: 'Right step five … praise them now' followed by '… er … great moustache!' and an uncomfortable silence following.

Ask 'What's the Worst that Could Happen and How?'

This is dynamic risk assessment in action and also looks to encourage discovered learning, which is the very best form of learning, since thinking of something and articulating it is an *active* process.

'Safety Watch' gives an example of a discussion around carrying a container of corrosive material. The observer was concerned that any cracks could lead to the leaking of this corrosive material or that dropping it would lead to fracturing the container and therefore splashing. This was intended as a lead-in to a discussion of the need for suitable PPE and the option of using carrying devices. However, the answer was: 'If I drop it, it may well explode.' So, a discussion that was apparently about PPE and manual handling became a *process* safety debate of great importance.

Clearly, this is just the very heart of a good discussion.

Question 'why?' of Any Unsafe Behaviour

As a passionate believer in the Just Culture perspective, I'd consider this element utterly vital and non-controversial – though also arguing it really shouldn't be just one thing towards the end of a list of *nine*.

Other models include the step 'Look for External Factors.' Specifically, it's suggested by exponents such as Scott Geller that a facilitator can inquire:

- Can the task be simplified to make it more user-friendly? He also suggests asking if unsafe behaviour is rewarded in any way, which, referring back to ABC analysis (speed and comfort are rewarding in themselves – see previous chapter), is simply another way of asking the same question.
- Are there basic barriers to safety, such as an individual being unaware of how to work safely, not having the right physical or psychological profile or the relevant equipment not being available?
- Is safe behaviour punished in any way (including peer-teasing as well as speed issues)?
- Is the law of unintended consequences applying? Because, for example, bonuses are driving corner cutting and under-reporting.

There are many excellent lists of questions of this type that are crammed full of useful prompts, but I'd like to get a little pedantic here to make a key point. Here's a direct quote from the Geller overview article in the Handbook of Occupational Safety and Workplace Health from 2015.

> A *systematic behaviour analysis of risky work practices can pinpoint many determinants of such behaviour, including inadequate management systems or supervisor behaviours that promote or inadvertently encourage at risk work.*

Absolutely. However, I suggest it's essential to substitute the *'can'* in the above sentence with *'will typically.'* If, based on Just Culture thinking and research, we see this determinant as a *probability,* rather than a *possibility,* and look for reasons to actively rule it out, then our analysis will be more accurate and the resulting actions more effectively targeted.

Such tinkering might provoke horror that I am daring to even mildly criticise the man who invented the term BBS and who has decades of successful consultancy behind him. However, I'm not suggesting a heretical revolution; but, I think, a simple but vital, nudge-inspired *evolution.*

Specifically, however, I *am* saying directly that I feel Krause, Geller, Daniels et al should, by now, have absorbed the thinking of Dekker, Reason, Cooper, and Hopkins. Because if they had, I think they would have fine-tuned and upgraded their BBS methodologies to think 'probable' rather than 'possible.' In his excellent BBS overview of 2015 (as above) from which the above quote comes, Geller references key papers by luminaries in the field of behaviourism or behavioural science such as Festinger, Zohar, Krause, Daniels, McSween, Skinner, and Deming (it's a very thorough article). But it makes no mention of 'Just Culture' or the hugely influential works on influencing skills by Thaller, Sunstein, and Cialdini. And as evidence of the validity of the charge that this American model of BBS is a bit insular and self-referencing, it's worth pointing out that his overview article references *himself* 29 times. Similarly, in Tom Krause's excellent and relatively recent *'Leading with Safety,'* Krause gives himself 16 references, while Kahneman gets only one and there are *none* for Reason, Dekker, Cooper, Conklin, or Hopkins. (So, I'm in very good company at least as I don't either!)

My issue is that even the best person-centered approach is, by definition, still *person*-centered. It's almost impossible to find a situation where there is nothing that can be *learned* and where empathy and pro-active analysis isn't appropriate, so unless we fluke it, the efficacy of our response is limited by the quality of our understanding. And perhaps most importantly, *starting with learning* is also often essential when we're considering workers who work peripatetically and *can't* be observed.

It should not be learning as a part of a classic observation and feedback approach, as many suggest, but observation (if used) as part of an analysis and learning based approach that pro-actively and directly enhances the environment. All methodologies and mindsets should flow from this.

At the end of his article, Geller raises several challenges for BBS going forward. Including:

- What do we do for sustainability over and above the observation and feedback process?
- How can we develop a 'brother's keeper' or interdependence culture, so that peer-to-peer feedback is supportive and corrective?
- How does the active involvement of management impact on a BBS process?

I argue that a European/Australian view has already addressed these issues. As well as the

above writers in 'B is for BBS' (March 2016 IOSH magazine), Bridget Leathley demonstrates a welcome focus on analysis, not coaching. She gives the example that if people are slipping on a wet floor, one should not just ask people to walk carefully. The approach should be to fix any leak and, if that's not possible, at the very least limit the number of people who walk on it and issue non-slip footwear. The top of the slips and trips hierarchy is to change the flooring. This is excellent 'hazard elimination' which, as Leathley rightly says, 'BBS is no substitute for.' (Though even this article stops short of mentioning a holistic approach including culture and leadership and many root causes of human such as fatigue and mental health).

To directly address Geller's challenges:

- What do we do for sustainability over and above the observation and feedback process? *(As above, a learning-based approach doesn't even have observation as the core methodology).*
- How can we develop a 'brother's keeper' (interdependence) culture, so that peer-to-peer feedback is supportive and corrective? *(By building a strong culture where challenging and being challenged, if required, is as stress free as possible. That's primarily a cultural issue where empathy and mutual respect are key, so start with a listening and learning methodology based on that).*
- How does the active involvement of management impact on a BBS process? *(They listen, they facilitate, and they praise BBS work. Sometimes they coach. Sometimes they need to challenge but they always play the central role in a holistic set of methodologies that is 'Cultural Safety').*

In short, again, we shouldn't see BBS methodologies as a stand-alone but as a key element of a holistic approach to cultural excellence. Process and personal safety are just overlapping subsets, as are health and wellbeing. So, the explicitly BBS element shouldn't primarily be peers challenging peers about unsafe behaviour; it is peers *conversing* with peers about the causes of and solutions to risk issues.

Ask What Corrective Action Is Required

Coaching and discovered learning skills are central to this element, as getting an individual to articulate what needs to be done is active for them. There is a loop back here to analysis, of course, and the DVD stresses that the observer 'might get an answer they weren't expecting.' In particular, they might well get a *'better* answer' than they themselves could have come up with. Again, I'd like to suggest that the underlying tone is one of a paternal expert pleasantly surprised by a workers' thoughts and knowledge. I argue the underlying tone should be that it shouldn't be a surprise – you're talking to *the* world expert on undertaking this task! So it's worth stressing again that, as the underlying tone is *everything,* it's vital to always work from the assumption that the person you are talking to is thoughtful and more knowledgeable than you can ever hope to be about the task in hand. (Unless, of course, they prove conclusively otherwise).

Specifically, it is suggested in the DVD that 'a training need might be identified.' This doffs a cap at Just Culture and blameless error. However, the nub of my concern about the approach comes through again as it points at a *training* need and that is a *person*-focused response. Coming at this from a Just Culture perspective, a training need *might* well be identified, but it's just as likely to be in *management*! Specifically, needing to address the way the task is set up, resourced, and supervised is, in most cases, a statistically far more likely answer.

We're in a minefield of nuances here, of course. I'm aware the distinctions I've just made may sound pedantic to some readers but I'm arguing they are *not* pedantic if you are genuinely coming at this discussion with Dekker's 'new view' or Conklin's 'pre-accident investigation' in mind. (Safety Differently coin this well: 'Assume the worker is the solution, not the problem.' Or, again, to quote the Poke Yoke management philosophy 'blame the process, not the person').

In addition, as the world of psychology (Kahneman et al) upgrades the estimated percentage of how much thinking is subconscious ever higher – latest estimates 90% plus I'd argue that the importance of nuance and subconscious mindset simply cannot be overstated.

Achieve a Commitment to Act ...

This very explicitly means *from the person being observed* because, of course, that's now the obvious close-out to a conversation of this type. It's what, I'm arguing, we were building to all along and this is really, for me, the clinching piece of evidence here as to why I think these approaches are too paternal. Because from a paternal perspective, this makes total sense. People are acting unsafely and they need to commit to stopping that, but I'm arguing this is squarely in 'we need to talk about your (bad) behaviour' territory.

We should be clear that we are there first and foremost to best find how we can facilitate safer behaviour, as it's more likely, fairer and more effective.

That being said – sometimes a commitment to act 'better' *will* be appropriate, of course! However, if 90% of the causes of unsafe behaviour are *environmental*, wouldn't simple logic dictate that 90% of these close-out commitments to act should come from the *observer*? (As in 'what do you need?').

Trusted leaders 'eat last' in Simon Sinek's terms and eliciting promises from workers not to repeat an unsafe act is the thing they should do last, if at all. I'd argue that the paternal parent mindset isn't where a world-class mindset sits. It needs to be adult-to-adult *always*. To explain: When we undertake surveys, one of the cultural clues we look for is *how* a program is referred to.

'Do you have time to do a SUSA with me?' is fine.
'I'm here to do a SUSA on you' is less fine.
'I'm going to SUSA you' is really not fine at all!

Once that later perception has taken hold, then it's efficacy as a methodology is inevitably limited. No one will give it a fair go, if nothing else. It's exactly like the cultural commentators who say you can always tell when morale is low in an organisation when employees stop saying 'we' and start saying 'them,' as in: *'They* need to sort this out.'

CONCLUSION TO WALK THROUGH OF 9 STEPS

It's worth remembering that the vast majority of these audits are undertaken *months,* if not *years* after the initial training course and the more opportunity there is for the wrong tone to be struck or the wrong emphasis given, the more often it will happen.

In the heat of battle, people will nearly always boil down a list of nine to a triptych of three and focus on the elements *they're most comfortable with.* If this is an organisation with a pathological compliance focus generally, that's almost certainly what they'll default to. If, in addition, the front-line managers have had limited or no soft skills training, that can be *really* asking for trouble and there are plenty of anecdotal examples showing that this is exactly what can result.

I'd like to instead suggest a five-stage model as detailed in an earlier book of mine *'Talking Safety.'*

SUMMARY OF RECOMMENDED 'WALK AND TALK' 5 STAGE MODEL

It's worth going back to basics and recalling again why we should devote valuable organisational time to undertake a walk-and-talk at all. It is never to try and catch someone out. Instead, it is to:

- Physically scan issues like housekeeping, while out and about;
- Learn something about the organisation and the people who work there through the eyes of those people. In particular to understand the unspoken but *vital* nuances;
- Understand why things have gone wrong or, if nothing wrong is seen, what people can be tempted to do.
- Model the behaviours and mindset we want the organisation to be built on;
- Coach, empower, facilitate, and even inspire your colleagues to discovered learning and embedded behaviours;
- Establish rapport, thus making more frequent and better-quality communications about safety more likely in the future.
    Occasionally, it is also to:
- Insist on an improvement in existing standards of behaviour and approach. In safety-critical situations this might need to be instant.

The following approach is designed to more directly map onto these aims. And remember, primarily, it's *quality* not *quantity* that we're after, so we don't have to stay out and about on site for a set amount of time. A good contact conversation is a good conversation.

INTRODUCE YOURSELF AND SET THE TONE

You're interested in them and their work and not trying to impress them with your knowledge, so, after some suitable small talk, you'll be instantly asking some of those excellent questions listed above and listening to the answers. You might use the FR and D of FORD model as well as the O. (Family, recreational interests, dreams and aspirations as well as occupation). So, do ask about their home life as well as the task they're undertaking and about process as well as personal safety issues. (As above). You should ask questions that show curiosity, commitment, and 'mindfulness' and never sound like you're there to catch them out. The assumption is that they are intelligent, committed, and knowledgeable. (Unless they prove conclusively otherwise).

Case Study

There's a head of safety of a specialist construction company who, nine times out of ten, *only* does this during her walk-and-talk. It's proving highly successful, as it covers most of the points in the list above but with the explicit aim at general learning and rapport-building. She finds that colleagues often raise safety issues spontaneously and when she has to address an issue more directly, her colleagues couldn't be more accommodating. (Indeed, she's one of those who respond to assertions that 'safety differently is new' with the comment 'been doing that all my career' and her 'door is always open' policy is one that doesn't have any element of forcefield about it!).

If you cover the following suggested questions thoroughly then you'll be bound to have a good discussion that may well generate some self-analysis in the person you're talking to.

Those are:

- What does this job involve?
- What can go wrong? (Include *process* safety issues).
- Why is that?
- What might happen if it did?
- How do we make sure that doesn't happen? ('What do you need?')
  And of course...
- Is there anything slow, uncomfortable or inconvenient about that?

Typically, as well as noticing the immediate risk once you've used your five whys analysis as above you may well find yourself considering such points as:

- Risk Assessments.
- Barriers.
- Signage.
- Supervision.
- Suitability of PPE.
- Training.
- Inductions.
- Selection and Monitoring of Contractors.
- Behaviours that are typical and not remarked on (norms).

Any of these issues worked through *systematically* and analytically and turned into an action plan is a long way from a basic hazard spot. But doing this systematically will nearly always require an in-depth conversation with the person doing the job. And that's where a dedicated open-minded safety contact comes in.

ANALYSIS

If you have seen anything worrying, or anything is raised, simply ask why, but ask it *curiously,* knowing that 90% of the time you'll get something interesting back that's learning about the organisation and environment. If there *is* an individual element, then 'curious why' will at least minimise defensiveness.

You can pro-actively start a why discussion, even if you've seen nothing or nothing has been raised, by, as above, asking the 'ABC analysis' temptation question: Is there's anything slow, uncomfortable, or inconvenient about doing this job safely. People will pretty much *always* have something to say here and, if they trust you, they'll tell you. Again, the conversation never has to move from the hypothetical and anonymous, so it's just as useful with peripatetic workers. The learning, however, *isn't* hypothetical.

One of the biggest problems we see with systems like a 'safety contact' is that analysis lacks depth. We often get asked how it is that things aren't getting any better, despite the fact that they've put straight lots of things that have been highlighted over the past year. The answer is that often auditors will list any number of problems but simply generate what's known as a 'crap list' of items that will recur. (The 'five whys' rule of thumb says if we keep asking why curiously on any answers, then we'll get to the root cause in five steps or less).

Five Whys Analysis

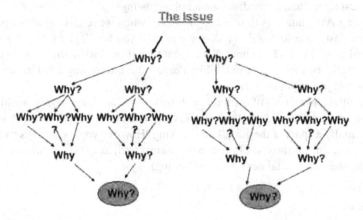

FIGURE 8.3

James Reason uses a mosquito analogy. He suggests that if you have a problem with mosquitoes, then a short-term solution is to buy nets, repellent, and swatters, but a better solution is to find the swamp they come from and *drain it at the source*.

One of the specific causes of these 'crap lists' is that people are often wary of approaching someone they don't know, especially if they're uncertain of the technicalities of the job. They therefore avoid talking to them at all and just pick on something visible and easy like housekeeping, then pop an action point down around that. A hazard may well be removed as a consequence and that's always a good thing but it's not necessarily the optimal use of time.

The VIP visit to the Macondo site the day before the explosion was accused of this – focusing on personal safety when process safety was, it transpired, key. An in-depth conversation following a comment 'I have to say I'm rather worried about this "it's just the bladder effect" explanation for this alarming looking reading' would have proved the most well-timed conversation in the history of safety.

THE TWO ELEPHANTS IN THE ROOM

This book has emphasised a holistic approach throughout so an excellent 'walk-and-talk' looking for personal safety issues should, if at all possible, also look for the two thumping great elephants in the room: ***Process safety*** and ***health***. The Deepwater example above illustrates and Andrew Hopkins says this well.

> *The behavioural approach is just as applicable to process safety as it is to personnel safety and auditing and monitoring activities should cover both. One of the criticisms leveled at behavioural safety generally is that it ignores process safety. It is vital that behavioural safety programmes face this challenge.*

Absolutely! And when we consider the numbers, everything that applies to process safety applies 100-fold to health issues. On 'walk-and-talks' we should of course talk about obvious risks such as falling down the stairs, but we should also talk about delayed but personally catastrophic risks relating to long-term health. (If you've jumped ahead to the 'applied' half of this book please do if you can see the chapters on wellbeing. It's the new frontier of BBS – indeed it makes the case that BBS is just a vital and immediate subset of wellbeing).

Keeping ABC analysis in mind is vital here when we're talking about potential exposure. Just because there's a risk that can't kill you for 20 years does not mean it *won't* kill you 20 years from now. Not a great thought as I write this as a 60-year-old hoping to still be enjoying grandchildren at 80. But utterly tragic if I'm only in my mid 20s and just about to start a family.

In addition, awareness of mental health issues has increased hugely worldwide in recent years and especially following COVID. Therefore, though it may have been asked initially as part of the 'small talk' asking 'How are you?' is always worth asking a second time at the end of the conversation. (And, as above, is covered in depth in the chapters on mental health and wellbeing).

COACHING

Ideally, you're already fully in coaching mode and modeling the mindset we want. If there is some specific learning you want to actively impart, however, you need to use questions to maximise discovered learning. More often, you'll just want to be using coaching techniques to lead them through a chain of thought. Coaching rather than telling is a key element of a strong and empowering culture and (again) was covered in earlier chapters. I'm hoping the 'feedback fish' resonates if nothing else! In summary, however, and to repeat myself slightly because it's so important, good coaching involves:

- Asking questions rather than telling striving to get them to say it first – proving they knew it and therefore maximising ownership of it.
- Maximise the praise opportunities that flow from this. (As appropriate to the situation).

The key issues here are that praise is about 20 times as effective in changing behaviour as criticism is. And secondly, if people say it first – even if both parties know they were led to it – studies of brain patterns show that the brain lights up and we 'own' it. This is utterly vital for issues of volition and internal promises.

Talking of promises.

PROMISES

If you've done the first three properly, then any promises required will be likely to have been made spontaneously and *internally* already, which is vital as this means they are far more likely to be *kept* when you're not around. This is hugely important when we're genuinely seeking to impact on what some call 'hearts and minds.'

There are times however, when we need a promise to be made genuinely because of a clear and present risk that can't be designed out in the near future or maybe cannot be designed out at all because of the cost. (For example, there should be scaffolding but it's very difficult to erect so we're having to use ladders. Or, we should upgrade the factory so we don't need PPE at all but we're moving to a new factory in 6 months and all cap ex is sunk in that. Or, of course: Ideally, we wouldn't have agreed to work to this timescale in this location – but somebody was going to as it's just hard rather than impossible to do it safely and the company really needed the work or jobs would have been lost).

A TOP TIP FROM INFLUENCING SKILLS – THE I's AND EYES HAVE IT!

As above, but worth repeating: Cialdini et al have many nudge-influenced tips for getting people to keep promises. (Though most are a variation on 'build a strong and trusting culture' as we're discussing here). Perhaps the most relevant for this part of a safety conversation, however, is the use of the 'I' word and eye contact. People who look you in the eye and say 'I will' are about five times less likely to break that promise than people who mumble at their feet.

If you ask someone to look after something important for you and they mumble 'Yeah, yeah, sure,' without catching your eye, you'd be a fool to trust them. By looking them square in the eye and asking 'Will you, please?' you maximise the chance of them looking back and answering 'Yes I will.' It's largely subconscious and they may well be unaware of those fingers they had crossed behind their back being uncrossed.

CLOSE OUT AND FOLLOW UP

Thank people sincerely for their time and insight. (See be positive and use praise above). Turn anything that needs doing into a SMART goal and then follow up and close out – or delegate a follow-up to ensure it was closed out and seek confirmation. In addition to the specific design benefits, this also shows genuine commitment.

Again, I'll keep stressing that Just Culture research shows that around 9 times out of 10, the person who needs to take an action away from the encounter is the one with the 'clipboard.'

We know that even small capital projects can be frustrating and time-consuming to set up and that we can all get so focused on the day-to-day realities that we're tempted to put long-term issues on the back burner. Basically, in the short term, it's a relief. (See ABC temptation above). However, it's vital that we don't file it in the back of the filing cabinet but actually stick to the SMART timetable and the person involved is returned to and updated on progress. If someone else is asked to talk to them, then don't cross it off the to-do list as 'delegated' until it's checked that they actually did.

FINALLY, *PLANNING* A 'WALK AND TALK'

If you fail to plan, you plan to fail, as the old saying goes. So what planning is required of a 'walk-and-talk?'

Firstly, we should check any 'walk-and-talk' databases or review previous safety contacts and consider:

- Who went on site last and what did they target?
- What did they find?
- What actions resulted? And how are things progressing?
  Also a basic
- Consideration of which jobs are being undertaken and when …

This will help ensure you don't turn up looking to target the same things as the last person. For example, you might note that no one has targeted working at height for a while so you could look to target that.

You could, of course, find yourself waiting *forever* for it to be totally convenient to undertake a safety contact audit, so don't be too considerate! However, please do show some discretion. If it's obvious that the people you need to talk to are *absolutely* flat-out busy, or if interrupting them could actually be dangerous, then give them some space and look at something else – even if only for a while. While you must make it clear you're not going to be fobbed off and leave, please *do* empathise and put yourself in their shoes. Simply ask yourself what would be reasonable if they were making an effort.

WALK AND TALK – SUMMARY

Even if you're not going to do anything directly with the workforce as part of the BBS strategy, then I'd strongly suggest that a good management 'walk-and-talk' is the one systemic process you really must install. In summary, executed well, a good 'walk-and-talk' will deliver benefits through:

- Learning, empowerment, engagement, and trust building!
  Also
- Less time and effort spent enforcing rules and regulations that are impractical or contradictory.
- Less admonishment of employees who are honestly trying their best and resent being (unfairly) 'told off,' with a consequent impact on discretionary effort, trust, and other 'psychological contract' issues.
- A reduction in 'crap lists,' that, while diligently circulated and rectified, simply recur time after time.

9 Desirable but *Not* Essential Methodologies

Measurement can certainly be helpful – giving the benefits of 'If you can measure it you can manage it' and 'what gets measured gets done.' But I'd like to argue here that if your methodology is driven by learning then it's desirable – and at times *highly* desirable – but *not* essential.

Behavioural data can be extremely useful for identifying areas of weakness and for tracking progress – both of which enhance learning. It can also be of great use for feedback as using data and illustration is a key element of coaching best practice. But it's only my *desirable* list, as doing it well is always difficult and in some cases impossible, such as with peripatetic workers. (If you're thinking it isn't always difficult to do well then please make sure to finish the chapter and read up on *accuracy and consistency assurance* methodologies and see if I've changed your mind!)

Certainly, I suggest there should always be an informed and thoughtful decision as to what measurement will look like and how it will be used. That's, again, if it is to be used at all. What we don't want is a thoughtless acceptance (or rejection) of a methodology because everyone else is doing it (or because it looks difficult) or because head office bought a really expensive bit of data tracking software and so everyone has to find data to enter into it ... regardless of how weak and misleading that data is!

LEAD AND BEHAVIOURAL MEASUREMENT

When running a research project in Manchester, we tried really hard to develop behavioural measurement as a science and were in due course commissioned by the UK Health and Safety Executive (HSE) to train a team of its inspectors in the techniques. Forgive a lack of modesty but I still haven't come across anything as advanced – though you be the judge. Regardless, I'll share those techniques here of course but before doing so it's worth making some predictions and observations:

- Lead measures (generally) often sound valid but in truth are vague and can suffer a disconnection with shop floor reality.
  However, despite that ...
- (Accurate) behavioural lead measures are the best lead measures as they enjoy the strongest direct correlation with risk and harm.
- How to collect good quality data isn't especially difficult to describe. The techniques, I hope, will sound sensible and achievable.

DOI: 10.1201/9781003449997-12

- Good quality behavioural data can also be used for goal-setting sessions as well as tracking and resource directing.

   But …
- Really good quality behavioural data is as rare as hen's teeth because of the resource requirement.

   And …
- This doesn't stop many an organisation entering poor quality data into very expensive computer programmes that generate lovely looking graphs and charts.

As the US civil war general's envoy famously said, pre-empting 'garbage in, garbage out' comments about computer analysis: 'The General accepts that the intelligence is unreliable and of poor quality, but keep supplying it please as he needs it for planning purposes.'

LEAD MEASURES

Context

Firstly, where does behavioural data fit within the general lead measure framework?

According to the extensive and free-to-use Campbell Institute website (www.thecampbellinstitute.org/research) lead measures should be *pro-active, predictable,* and *preventative*. Other adjectives often required of them, it says, include actionable, achievable, explainable, meaningful, timely, transparent, useful, and valid. I'm not sure whether such a list scores most highly for thorough and useful or for overlapping and intimidating!

Regardless, three broad sets can be identified:

Operations such as risk assessment, compliance, corrective action close-out, change management, and training.

Systems would include permit to work, surveys, disciplinary issues, risk, and hazard analysis.

Behaviour (our primary focus here) including – as well as PPE compliance and man machine interaction etc, culture, leadership, front-line engagement, 'walk-and-talk' and other 'visible' methodologies.

I'd like to pick some operations and systems examples to illustrate the observation that, despite some very sexy-looking graphics, they often tell us little of use.

For example, in an improving culture, hazard-spotting metrics increase and it's well documented that first aid and even injury rates often increase as the culture improves. Not great for tracking purposes and in 'art not science' territory.

A less obvious example would be a training metric like: 'Percentage of employees trained in safety X compared to the percentage of employees we intended to train.' If we halve the number missed last year – from say 40% missed to 20% missed we can congratulate ourselves on our 50% reduction 'step change' achievement. More than that, it may well be a fun course, delivered with passion by entertaining trainers, so the happy sheets will look great too. We certainly now have several impressive looking charts to wave at the C-suite and visiting auditors.

However, several questions remain:

- Was the training based on an accurate gap analysis of need? Or was it just considered useful by one individual? Or something that used up budget to ensure next year's isn't cut? Or was it always intended as a 'tick box' exercise and something to just wave at a regulator or client?
- What percentage of the at-risk population were targeted for training? All of them or just some? Are contractors – who might well do the most difficult and risky work – included?
- Were there any language or comprehension issues? For example: 'I am new here, the trainer was nice and so I gave him a good score even though I didn't understand much.'
- Were the messages clear and sticky so that 'what to do' is remembered several months later?
- Was '*Why* we need you to do this more/less' covered in enough depth for the delegate to demonstrate some 'operational dexterity' should the situation not be clear-cut?
- Were the new behaviours requested practiced to the extent that individuals left the course feeling confident enough to attempt them in action when given the chance? It's worth stressing that even highly-skilled and engaging trainers role-playing at the front, so that delegates can watch and say: 'Yes, I see what you mean,' isn't the same thing as actually *practicing* it.
- Were the behaviours requested followed up by the organisation so that opportunities to use them draw negative feedback if missed and positive feedback if taken?
- Was the training followed up by line managers and coaches so that early attempts to use new skills could be discussed and any support required could be actioned?
- Did we follow up to ensure that the behaviours happening more/less often deliver something unambiguously positive for the organisation? Or are they just delivering something different, or, even something worse, because of an 'unintended consequence?' (Perhaps the most infamous unintended safety consequence is when a 'target zero' approach drives reporting underground because people are scared to 'let the team down.' Another infamous example: In the year following '9/11' there was an increase in fatalities on the roads on the USA that actually *exceeded* the number killed on the day in question as so many people switched away from flying to the relatively far more dangerous driving).

Even the above isn't an exhaustive list, but it does make the point that simply asking an employee 'have you been trained?' and 'how was the course?' and getting the responses 'yes' and 'great, actually' doesn't *necessarily* prove much. Here's another example from a large nationwide company that measured whether or not quotas were hit for walk and talks. They tracked the number of interactions undertaken but (despite our urging) never cross-referenced that with the *quality* of those (Figure 9.1).

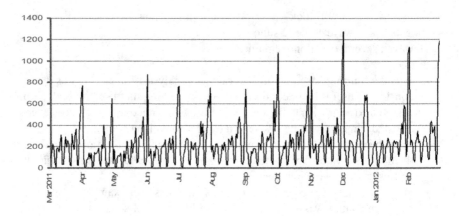

FIGURE 9.1

Since I have no data, I don't know for certain that lots of these interactions were knocked off in a rush and of poor quality. (But anecdotal evidence was that some were actually entirely *made up*).

In short, many a pie chart based on 'lead measure' data looks convincing and face-valid but the simple question that underpins all good science must be rigorously asked and researched. Is this *causation* we are looking at or merely *correlation*? And if causation is claimed, then how can we *prove* it's the case.

CULTURE AND CLIMATE SURVEYS

Surveys can be a good way to pro-actively find out what's going on.

My experience, however, is that the tick-box survey, especially if it's online, suffers from two huge problems. Firstly, the person filling it in probably cares about it far less than the people who wrote it. This almost certainly means a certain amount of random ticking and hence error variance. Secondly, the typical respondent will not believe for one minute that the survey is anonymous and therefore may well second-guess the answers they think they ought to best supply. (Again, this is especially true of electronic surveys).

Combined, this may well result in huge amounts of error variance, meaning that a mean score of 50 (plus or minus 15) actually needs to increase to a mean score of at least 80 (plus or minus 15) before we can be *scientifically certain* that things have genuinely improved.

Face-to-face culture surveys are much better as the respondents, who are usually in groups, can be assured personally that the pollster doesn't want to know their names, there are no right or wrong answers and the purpose is just to find out what they *think*, why they think it and to hear some examples that illustrate. This provides really rich data and a clear opportunity for an experienced researcher to spot any 'best foot forward' responses and challenge them by asking for examples. (A hybrid really useful, especially in large companies, is to seek the best of both worlds by using an online survey initially, then do face-to-face anonymous 'deep dives' into interesting findings).

Here is a typical example of our own culture survey, with the management scores above and workforce scores below. In *both* cases, the groups seemed motivated and honest, so it illustrates the need for a proper sampling methodology that covers all elements of the organisation!

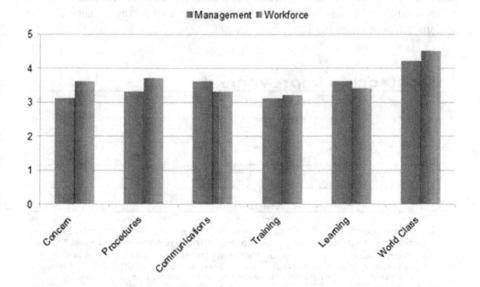

FIGURE 9.2

We've used a generic illustration here to protect client confidentiality (Figure 9.2). However, typically you'll find that management claim to be 'not too bad actually and are particularly strong in the areas of analysis and communication.' While the workers are saying *'you're got to be joking!'*

However, unless the discrepancy is so great as to show that management *is* utterly deluded, it matters not that scores diverge. What's important is *what we learn from them*. Indeed, exactly why management and workers disagree is merely one of a number of hugely fruitful learning opportunities.

The symbolism of conducting a survey has an impact too under the heading 'What gets measured gets done' because it tells the workforce what we're interested in. A caveat to add, however, is that how well we measure it and how well it appears we are measuring it impacts on that. If it is a valid measure and if it is also a face-valid measure, then it's well worth doing. (Unless we don't do anything with the findings, in which case the unintended consequences that kick in – enhanced cynicism usually – leaves us knowing we'd have been better off doing nothing!)

However, caveats aside, if we do effectively apply the learning from such data, we will impact significantly on front-line risk. So, this is therefore, I'd argue, potentially a valid element of a holistic 'behavioural safety' approach.

CASE STUDY

Here's an interesting case study demonstrating the power of what gets measured gets done. Examples above have illustrated the principle that 'we get what we demonstrate we want' and taking the trouble to measure it demonstrates that want. This example, from an Italian manufacturing plant, adds an amusing refinement to the famous 'what gets measured gets done' truism.

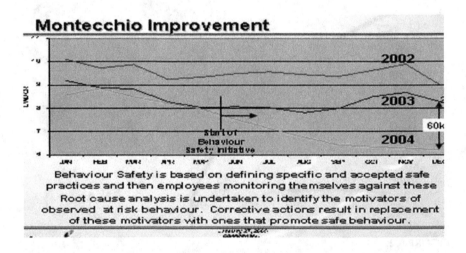

FIGURE 9.3 Preparing for an audit.

You will note that the scores start to improve not when we undertook the survey and gave feedback and suggestions but on the day the site was told 'we're coming to measure behaviour.' So it's also 'what's *about to be* measured gets done.'

BEHAVIOURAL MEASURES

This final section is about out-and-out 'behavioural measurement' as it's typically understood. I really hope you'll find that this section, even as a sub chapter, is at least as in-depth a coverage of behavioural measurement as anything available in the literature.

It's worth re-stating again: What's especially good about front-line behavioural measures are their point-to-point correspondence with risk outcome. Here are two case study examples.

CASE STUDY 1 – MANUFACTURER

The first chart shows the mean scores for a variety of categories: Housekeeping, PPE, worker/machine interface etc, which caused the vast majority of incidents for this client. There were a variety of easy wins to be had here. For example, issues around plant pedestrian interface RE the pedestrian door that would remain locked most of the day unless you found the key holder and asked them to open it.

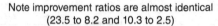

Behavioural Scores and LTI Rate

Note improvement ratios are almost identical
(23.5 to 8.2 and 10.3 to 2.5)

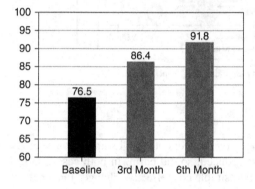
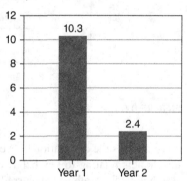

FIGURE 9.4

You can tell from the incident rate, that this was a very rough and ready industry with a lot of challenges, but management commitment was high. The point I'd like to make here is that the ratio of scores going up and accidents coming down is very closely correlated, so it's a reasonable assumption that there's no need for any cross-referencing and caveats as with the training pie charts above. Improvement, initially at least, was being measured *accurately*.

CASE STUDY 2 – CHEMICAL MANUFACTURER

This second example is even simpler as it involves just one simple behaviour and just one very cheap solution. At this plant, the vast majority of days lost were to burns and dermatological issues because there were lots of instances when workers would need to manually handle containers of caustic soda, which is all this site manufactured. Sometimes, they'd not wear gloves when they did this and sometimes the containers leaked a little.

We found that only one size of glove was provided because it was cheaper than supplying a variety of sizes and no one had ever complained. People with big hands couldn't get the gloves on and people with small hands found the fingers flopped about in an inconvenient way. The solution – to provide three sizes of glove – was cheap and its impact on days lost (their favoured lagging measure of safety) was dramatic.

ITALIAN MANUFACTURING PLANT

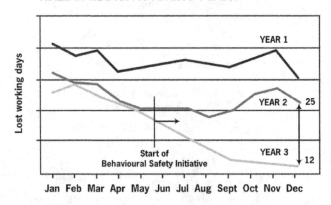

FIGURE 9.5

Again, it's the correlation between the two sets of data that I want to consider. As unsafe acts relating to key behaviours reduced, so did harm. In both cases, understanding why unsafe acts occurred led to simple cost-effective changes that enabled an improvement in performance.

ACCURATE BEHAVIOURAL MEASUREMENT

Here's another example of 'if we can measure it we can manage it.' It's from a utility company which had thousands of walkways across dozens of sites. Many were degraded and hazardous but efforts to improve matters were piece-meal. Often contractors couldn't be bothered to quote for small jobs and the paperwork required off-putting. In this case, we were provided with a team of BBS assessors who toured the whole site scoring each 40 yards of walkway against agreed standards who came back with thousands of data points. The feedback session to senior management proved both interesting and productive as the score was so low as to be alarming. The head of legal was asked 'are we liable if someone trips on one of those walkways?' and when the answer was 'yes' the company quickly found a contractor who could quote for a large company-wide job and several months later the walkways all looked like the second picture (Figure 9.6).

FIGURE 9.6

BEHAVIOURAL MEASUREMENT – A SUMMARY AND CHALLENGE

Many practitioners, like BST describe the classic BBS model as:

- Identify the key behaviours.
- Measure them.
- Feedback the results.
- Remove the barriers.

Clearly, if we base our approach on this model, then accurate measurement is key, as it is for Scott Geller's 'DO IT' approach which stresses 'I' is for intervene and 'T' is for 'test that the intervention worked.'

Specifically, accurate measurement allows:

- Benchmarking against other organisations/other parts of the organisation/ other shifts.
- Meaningful goal-setting sessions to describe accurately 'where we are now' and define clearly 'where we want to be.'
- The impact of training, design solutions and other initiatives and processes to be objectively assessed.

All things considered, it's clear that *accurate* measurement of front-line behaviours is a good thing and, if viable, better than no measurement. My experience, however, is that achieving accuracy in the medium to long term is, as stated above, far more difficult than most people realise.

Obviously, I need to justify that sweeping statement …

Imagine two observers tasked to go out with a check-sheet asking to look for behaviours relating to PPE, housekeeping, the interaction of people and machines and movement about the site. This is because we know from talking to employees that if we can crack these items then we have covered 80% plus of all lost-time incident causes. Consider that they are both sensible, experienced, diligent, and motivated. (Though, in truth, this can be something of an assumption in places where ownership of and/or the perceived credibility of a BBS approach is poor).

Observers A and B walk around the site scoring what they see, diligently, and put their observations into a spreadsheet where the week's data is collated and turned into a sexy-looking pie chart or bar graph. Here is a list of the things that can put error variance into the behavioural data system because even diligent, experienced people may well see two different things. In short, what A thinks is common sense and what B thinks is common sense may differ wildly.

Classic problems with checklists that generate error variance:

THERE ARE VAGUE ITEM DEFINITIONS

'Make sure people are wearing overalls and wearing them correctly,' for example.

If the question: 'What exactly do you mean by that' can be asked then the definition is too vague. For example, here you may need to say that they:

- Are done-up to the neck.
- Have sleeves rolled down and outside of gauntlet gloves.
- Have trousers outside of boots.
- Nips in the fabric are allowed up to 2 cm but no tears that could snag or let in chemical splashes.
- They must not be overly (*) impregnated with oily waste.

PHOTOGRAPHS ARE REQUIRED

Where it's virtually impossible to define 'overly' as above* in words, we need **borderline** photographs to illustrate where acceptable falls. Importantly, these should *not* be good and bad examples that mean 'picky' people will use the good photographs as the standard to work from, as 'anything worse than that fails,' while more laissez-fair types will gravitate towards the bad example on the basis that 'anything better than that is OK.'

GENERIC RULES ARE NEEDED

Two observers could watch a fork-lift truck driver working hard for two minutes and see dozens of individual movements, many of which might be borderline in terms of being acceptable or unacceptable. Was that too quick a turn? Was that pallet engagement a bit too aggressive? Was that head turn sufficient? One observer, scoring these harshly, and one scoring leniently could come up with wildly diverging scores. A generic rule, while it will of course reduce sensitivity, *it will greatly enhance consistency while remaining sensitive enough.*

For example:

'Watch the driver for two minutes, if they transgress any of the following key points fail them, if not pass them.' Then go and watch five other drivers and come back with 6 marks in total. (A similar rule might be applied to any rapid-fire frequent behaviour such as the use of hand operated tools or manual work on a production line).

MAPS FOR HOUSEKEEPING ITEMS

Maps are always useful for telling observers how many housekeeping marks to give in a specific area. For example, make clear that an area (a room or a designated part of a big room) gets one mark and which other sections get another to ensure that one person doesn't score every pile of boxes and access to every fire extinguisher and another person just the room. Again, *consistency* is key.

Walkways can, for example, be scored as 'circa 30 m long,' so a standard 25 m walkway gets one mark and a 100 m walkway three or four marks, depending on the location of user-friendly 'border points' like doors, side paths or drainpipes!

PEOPLE FORGET TO SCORE THE SAFES

It might be that even though all data is collated and turned into an overall percentage, one observer, for example, watches 10 scaffolders access a build and (correctly) scores it as 'eight safe and two unsafe.' Or 80% compliance. However, we know that unsafe actions leap out at us in these circumstances, so a second observer, only 'seeing' these two miscreants taking a shortcut access to the site only, scores it 'two unsafe.' (Or 0% compliance!)

A LACK OF TECHNICAL KNOWLEDGE

A subset of this error-inducing issue is where an observer is watching something technical. Whether a worker has a hard hat on or not is easy to score, as is whether employees are holding the handrail or not. But – items relating to the use of specialist tools or lifting operations on such as a drill floor are harder, especially, if for example, you're a *cook* by training! And one of our very best BBS team volunteers ever was a cook on an oil rig. His team mate was a medic. (Hello Mark R and Graham M).

We've often found that observers don't feel comfortable 'judging' colleagues when they're a bit uncertain and often then simply decline to give a mark either way. What's needed here is some cross-function peer-to-peer coaching in what to look for.

It was during one such session that one of my favorite observer training stories occurred. Being shown around a kitchen, a driller was shown an emergency stop button. He took the cook up to the drill floor and, paraphrasing the 'that's not a knife; *this* is a knife' quote in the film *Crocodile Dundee,* said 'What you showed me isn't an emergency response button; *this* is an emergency response button,' pointing out something rather larger and more user-friendly.

POOR SAMPLING

We know very well how to sample accurately as around the world exit polls accurately predict an election result minutes after a poll has closed. Regardless, the sampling 'Catch 22' can often apply.

The trouble is that, like water always taking the easiest route down a hill, observers will, left to their own devices, get into a user-friendly pattern and we'll get scores that will, for example, tell us how safe the site is just after lunch on a Tuesday when it's nice and quiet and safe – and most convenient for taking an observation. We, of course, most need to take observations when it's usually the *least* convenient time to do so as people are stretched and busy as that's when gaps in the system so often announce themselves.

This needs planning effort and management commitment. An example of this would be the client that used random number tables to ensure that operatives could never predict when and where an observer would appear and was much amused when one observer drew 10 am and 11 am, both clockwise, on Monday morning as their two observations for the week!

ACCURACY AND CONSISTENCY CHECKS

In the Manchester project, we tried very hard to design out the faults described above by, for example, feeling it right to sacrifice a little sensitivity for the consistency that robust 'generic rules' brought. In order to check how well we were doing with this, we designed the 'accuracy and consistency' methodology and, as above, the UK HSE was impressed enough to ask us to train a team of officers in the techniques to be utilised in a push in the building industry. (It didn't last long as a project because of manpower issues mainly but we'd already found that operatives did not like being observed by HSE employees! – this decades ago now though when there was an awful lot more to observe on a building site!)

What we found was that achieving 'really good data' was indeed possible but almost no organisation had the appetite to collect high-quality comparison data over any sort of meaningful amount of time. (It's widely agreed that embedded culture change takes several years at least. Tracking data that *proves* a meaningful improvement therefore needs to remain accurate for this time).

We found that sometimes inconsistencies evened themselves out so that two different observers could come back with a similar percentage score. However, if this didn't also reflect a similar *number* of scores, then that agreement would be highly unlikely to be repeated. We found that if a simple cross-check with agreement for both the percentage score *and* number of observations made were both above a co-efficient of 0.9, then the data was solid. If either of these fell below 0.8, we felt that was too much error variance. (I am a social scientist and once passed a 3-hour exam based entirely on the study of Rosenthal and Rosnow's 600 page 'Essentials of Behavioural Research: Methods and Data Analysis' but need to admit here that the figures the research team agreed are as 'rough and ready' as they look and not subject to t-tests and co-efficient analysis!)

The basic A&C methodology was for two or more, observers to go out and score the same site without any sort of communication. By communication, we meant eye contact, nods, pointing, winks, thumbs up and other 'what do you think?' non-verbal communications.

Here's a real-life example based on the above 'possible problems' text:

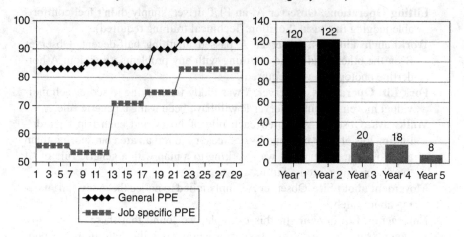

Safety Performance
(Behavioural Measures and Working Days Lost)

Accuracy and Consistency Data

	A			B	
	Safe	Unsafe		Safe	Unsafe
1. Lifting Operations/ Use of Tools	4	0		0	0
2. Workbench tidiness	10	1		1	10
3. Fork Lift Truck Operations	2	2		24	2
4. Walkways	23	48		1	17
5. Movement Around Site	31	2		0	2
6. Walkways and emergency equipment	15	5		19	1

59% (143)   58.5% (77)

FIGURE 9.7

As you can see, the similarity in *percentage* is fine, scoring a co-efficient of 0.99 (58.5 divided by 59). However, it's highly unlikely to be repeated as the co-efficient score for the *number* of marks given is only 0.56 (43 divided by 77). The method from here is simply to 'eyeball' the data – which is honestly a technical term meaning taking an objective overview of what sense it makes – identify where the discrepancies are, discuss why they occurred, agree on a remedy, then of course re-run the A&C check to make sure the remedy actually works.

The problems we found in order:

Lifting Operations. Observer A, an FLT driver, simply didn't feel comfortable judging this task. (Solution: Technical training required).

Workbench Tidiness. Observer A passed anything half-decent, Observer B on the other hand failed anything with any problems. (Solution: A borderline photograph was required).

Fork Lift Operations. Observer A was giddy from trying to score each individual movement but Observer B will have been using a generic rule.

Walkways. Observer A scored each pile of boxes and each trip hazard – finding lots of 'safe bits' and lots of reasons to fail an area but scoring them all. However, Observer B was working to a map with a generic rule of 'if there's anything wrong, fail it.'

Movement about Site. Observer A simply failed to notice the many staff moving about safely.

Emergency Equipment. In this example, we found that there were maps, generic rules, and example photographs in use, but the definitions and photographs weren't as pinpoint as they could be and one observer had started to leave them behind and 'use his judgement.' You can see that this resulted in one observer scored this 95% safe and the other only 75%. (Crucially, this amount of variation isn't much use regarding a goal-setting session which starts with the observation 'Overall we're at 80%; we hope to halve that and get to 90% in the next three months').

Three observations:

1. I don't think that over 30 years we've ever had a client that could point to A&C scores that passed muster for more than three or four months.
2. This didn't stop many from turning observation data into sexy-looking pie and bar charts and using these in meetings with no caveats. Many also used the data in goal-setting sessions and on feedback charts over a period of years and included summaries of them on 'dashboards.'
3. This also didn't stop some clients investing in very expensive computer tracking programmes. We even had one we provided ourselves. It was reasonably priced by comparison to others but, even so, because of points one and two, we took an executive decision to stop selling it.

BUT ... MERELY BEHAVIOURAL BENCHMARKING IS RELATIVELY EASY

All this aside, however, being able to accurately compare different clients/different factories or regions or to track the effectiveness of processes and initiatives over time remains extremely useful. A robustly designed and quality-controlled list of (say) the six most key items benchmarked against collated means and norms always proves interesting and useful to clients. Just use the methodologies above to make sure the data is genuinely comparable!

This is especially so when a BBS intervention based on *analysis* and *empowerment* – described above as the essential elements – has been implemented. Re-doing the bench-marking is an excellent way of tracking progress and proving effectiveness. For example, if FLT maneuvers have improved from say 50% to 75%, then – regardless of sensitivity lost because of generic rules – I guarantee that half as many crash barriers will be scratched and dented! (And the risk to pedestrians will also have halved even if near miss reporting is actually up because the culture has improved).

I hope you now have everything you need to generate accurate data on behaviours in-house. Even if 'snapshot' data is all that is viable …

GOAL-SETTING AND FEEDBACK SESSIONS

If the data collected is accurate enough to be credible (at least), then as well as feed-back charts and the like, a goal-setting session is possible. Ideally, this will be run by the workforce themselves with colleagues assembled in a canteen or similar to debate and agree on a hard but realistic goal. Following this, charts can be updated weekly to track progress towards this goal. (Having outlined a high degree of scepti-cism about the accuracy of such *weekly* updates a clearly defined and resourced and keenly targeted benchmarking exercise over a period of months does allow for a follow up feedback session).

A well-attended and motivated session like this can be a hugely-important event in the BBS process. This is especially so when someone astutely observes 'We can get from X to Y in two months, no problem, but we'll need Z to achieve it.' For exam-ple, this occurred quite spectacularly at a factory in Manchester when we discussed the bypassing of interlock gates. The workforce was genuinely stunned at how bad things had got. (Around 15% compliance). After an excellent discussion with some astute observations and suggestions, a target much higher than we were hoping for was agreed (we hoped for 60% as an initial target but '90% minimum' was suggested and agreed by all). It was achieved the *following week*. That's the way to do it!

(Again, and for the last time, we could merely have pro-actively asked workers to describe a problem that fell under the 'safety as said and safety as done' heading. Understood the reason why anonymous people were tempted to cut corners and addressed to the issue that only one person had a key and was often difficult to find …).

Things that can go wrong with this approach:

• The chart isn't updated promptly every Monday morning as promised but can stay unchanged for weeks on end. The unintended consequence is that this now becomes a well-placed and colorful reminder for all of the luke-warm commitment to the process by management.
• The session is led by a notoriously authoritarian management and the attendees agree with anything suggested but don't actively participate in any meaningful way. Thus, an opportunity to engage the workforce in the process is lost.

- The session *is* run in a coaching style and by peers or management, but clumsily. Therefore, a response to a 'surely we can do better than that?' prompt is something like: 'Look, why don't we just cut to the chase and you tell us what target you want us to agree to. Then we can all clear off and get back to work.'

Very explicitly, I'm suggesting some basic training in the generic interpersonal skills – presentation skills especially – for all involved in such a process. And again, these basic skills really help with that all important 'operational dexterity' we want to see in our volunteers as obstacles arise in the process. Heckles from goal-setting session attendees included.

CASE STUDY

I'd like to include a case study with cultural measurement at its core that's in line with the observation 'if it systematically reduces front line risk then it's a variation on BBS.'

Some years ago, I had a client with a CEO who clearly had a firm grasp of the concepts discussed above. He commissioned an extensive and international safety culture survey which took a considerable amount of time, travel, and effort. When the data was collected and collated, we discussed which categorisations he was most interested in. By Country? Division? Department? …

It turned out to be *none* of those. Essentially, he just wanted one simple 'external expert' generated overall score – though already broadly knowing what it would be. He simply required the risk implications of having a merely average culture spelt out to his board in simple and direct terms. He felt (correctly) that they'd convinced themselves that they were 'pretty good all things considered.' After all, it had been years since the latest fatality and, despite the nature of the industry, serious injuries were reasonably infrequent. He sensed that, in short, his colleagues slept well at night but on his travels, he'd seen excellence for himself and *he didn't*.

Following the session, a thorough strategy was devised, costed, funded, and implemented. I argue that the CEOs strategic vision and political skill here was an excellent piece of BBS.

Also excellent BBS:

- A supervisor saying 'sorry I didn't mean to say 'safely but by Friday… I meant safely *and* by Friday but if you don't think that's viable let's grab a table, a coke and a mars bar and talk it through.' Because creating a strong, open, learning focused culture based on respectful and mindful dialogue is excellent BBS.
- Effective design work – as high up the safety hierarchy as possible – that designs the risk out entirely is excellent BBS.
- Providing good quality PPE that is far more comfortable to wear than original kit is excellent BBS.
- Coaching safe behaviour so that the 'why' it's required sticks internally is excellent BBS.

- 'Catching' someone doing something safely and praising them (especially if the safe way requires some effort that we can't design out) is excellent BBS.
- Always leading by example (in a good way!) is excellent BBS.
- Stopping to tidy up a potential trip hazard is excellent BBS.
- Systemically reducing the amount of fatigue or mental health issues suffered by the workforce so fewer workers are distracted and/or fatalistic and/or have bad interactions or make bad decisions is excellent BBS.

You get the idea, I'm sure. Anything at all that helps your company to reduce the number of employees exposed to risk *on the front line* – especially in the medium to long term – is, I'm arguing, excellent BBS. Don't let anyone tell you otherwise and *especially* if their bonus is related to sales volume!

One size does not fit all and different organisations and cultures mean that every holistic programme needs tailoring. However, infinitely varied though people's behaviour is – it's often quite easy to predict and therefore quite easy to address pro-actively.

Bibliography

Ajzen, I. (1991). The theory of planned behaviour. *Organizational Behaviour and Human Decision Processes*, 50(2). 179–211

Beer, M. et al. (2016). *Why Leadership Training Fails – and What to Do About It*. Harvard Business Review.

Blanchard, K. and Johnson, S. (1982). *The One Minute Manager*. William Morrow.

Buckingham, M. and Clifton, D.A. (2005). *Now, Discover Your Strengths*. Pocket Books.

Cialdini, R. et al. (2017). *Yes, 50 Secrets from the Science of Persuasion*. Profile Books.

CIPD and Mind. People manager's guide to mental health. Bit.ly/CIPD-mind-mental-health.

Conklin, T. (2012). *The Pre-Accident Investigation: An Introduction to organizational Safety*. CRC Press.

Cooper, M.D. (2009). *Behavioural Safety: A Framework for Success*. BSMS.

Covey, S.R. (2004). *The 7 Habits of Highly Effective People*. Simon & Schuster.

Daniels, A.C. (2007). *Other Peoples Habits: How to Use Positive Reinforcement to Bring Out the Best in People Around You*. Performance Management Publications.

Daniels, A. and Agnew, J. (2010). *Safe by Accident? Take the Luck Out of Safety*. Performance Management Publications.

Dekker, S. (2008). *The Field Guide to Understanding Human Error*. Ashgate.

Dekker, S. (2007). *Just Culture: Balancing Safety and Accountability*. Ashgate.

Dekker, S. (2014). *Safety Differently: Human Factors for a New Era*. CRC Press.

Deming, W.E. (1986). *Out of the Crisis*. MIT Press.

Dweck, C. (2008). *Mindset: The New Psychology of Success*. Random House.

Edmonson, A. (2019). *The Fearless Organization: Creating Psychological Safety in the Workplace for Learning, Innovation, and Growth*. Wiley.

Frankl, V.E. (1959). *Man's Search for Meaning*. Riderbooks.

Geller, E.S. (2001). *The Psychology of Safety Handbook*. Lewis.

Geller, E.S. and Robinson, Z.J. (2015). Behavior based approaches to occupational safety. In *The Wiley Handbook of Occupational Safety and Workplace Health*. Wiley Blackwell.

Harris, T. (1995). *'I'm OK, You're OK*. Arrow.

Heinrich, H.W. (1959). *Industrial Accident Prevention: A Scientific Approach* (4th ed.). McGraw-Hill.

Hersey, P. and Blanchard, K. (1969). Life cycle theory of leadership. *Training and Development Journal*. 23. 26–34.

Hilton, M. and Whiteford, H. (2010). Association between psychological distress, workplace accidents, workplace failure and workplace successes. *International Archives of Occupational and Environmental Health*. 83(8). 923–933.

Hopkins, A. (2008). *Failure to Learn: The BP Texas City Refinery Disaster*. CCH Australia.

Hopkins, A. (2012). *Disastrous Decisions: The Human and organizational Causes of the Gulf of Mexico Blowout*. CCH Australia.

Joseph Rowntree Foundation. (2022). UK Poverty.

Kahneman, D. (2011). *Thinking Fast and Slow*. Farrar, Straus and Giroux.

Kotter, J.P. (1996). *Leading Change*. Harvard Business School Press.

Krause, T.R. and Hindley, J.H. (1990). *The Behaviour Based Safety Process: Managing Involvement for an Injury-Free Culture*. Van Nostrand Reinhold.

Krause, T. (2005). *Leading with Safety*. John Wiley & Sons.

Leathley, B. (2016). *B Is for Behavioural Safety*. Marsh: IOSH Magazine.

Marsh, T. (2021). *Talking Health, Safety and Wellbeing: Building an Empowering Culture in a Post-COVID World*. Routledge.

Maslow, A.H. (1971). *The Farther Reaches of Human Nature*. The Viking Press.

McEnroe, J. (2003). *Serious*. Time-Warner.

Morgan, G. (2006). *Images of Organisations*. Sage.

Omand, D. (2020). *How Spies Think*. Penguin.

ONS suicide statistics, bit.ly/ONS-suicides-england-wales.

Pavlov, I.P. (1927). *Conditioned Reflexes: An Investigation of the Physiological Activity of the Cerebral Cortex*. Oxford University Press.

Peterson, D. (2019). *12 Rules for Life: An Antidote to Chaos*. Penguin.

Reason, J. (1997). *Managing the Risk of Organisational Accidents*. Ashgate.

Reason, J. (2008). *The Human Contribution: Unsafe Acts, Accidents and Heroic Recoveries*. Ashgate.

Rosenthal, R. and Rosnow, R. (1991). *Essentials of Behavioral Research: Methods and Data Analysis*. McGraw-Hill.

Ross, L. (1977). The intuitive psychologist and his shortcomings: Distortions in the attribution process. In L. Berkowitz (Ed.), *Advances in Experimental Social Psychology*. Academic Press

Rushton, L. et al. (2012). *The Burden of Occupational Cancer in the UK*. RR931. HSE.

Schneider, B. (1987). The people make the place. *Personnel Psychology*, 40. 437–453.

Sinek, S. (2017). *Leaders Eat Last*. Penguin.

Skinner, B. (1974). *About Behaviourism*. Knopf.

Syed, M. (2015). *Black Box Thinking*. John Murray.

Thaler, R. and Sunstein, C. (2009). *Nudge: Improving Decisions About Health, Wealth, and Happiness*. Penguin.

Vroom, V. (1992). *Management and Motivation*. Penguin.

Warr, P. (2009). Environmental "vitamins", personal judgments, work values, and happiness. In: S. Cartwright and C. L. Cooper (Eds.), *The Oxford Handbook of Organizational Well-Being*. Oxford University Press.

Wilkinson, R. and Pickett, K. (2011). *The Spirit Level: Why Equality Is Better for Everyone*. Penguin.

Index

Page numerals in *italics* refer to figures and those in **bold** refer to tables

Printed in the United States
by Baker & Taylor Publisher Services